BROKEN SWASTIKA
The Defeat of the Luftwaffe

BROKEN SWASTIKA
The Defeat of the Luftwaffe

by

WERNER BAUMBACH

Translated by
FREDERICK HOLT

ILLUSTRATED

London
ROBERT HALE LIMITED
63 Old Brompton Road S.W.7

PRINTED IN GREAT BRITAIN
BY EBENEZER BAYLIS AND SON, LIMITED
THE TRINITY PRESS, WORCESTER, AND LONDON

CONTENTS

ILLUSTRATIONS

Note by the German Publishers

The publishers are fully conscious of their responsibility in producing this book. The air war, and everything associated with it which is described in this book, can be a provocative subject for militant minds. The risk must be taken if we are to know the true course of events which brought our nation to the political, military, economic and moral catastrophe of 1945. The reader cannot avoid gazing into the abyss. No one with a real sense of responsibility who goes deeply into all this can escape the conclusion that there can be no forgetting some of the things that happened in the years 1939 to 1945. Hitler's name, the NSDAP[1], the Gestapo, the S.S., and, unhappily, much that the German Wehrmacht did, will for ever fill us with shame. But that does not prevent the truth being told or truth and honour being separated from lies and fraud. Foreign countries have done justice to Baumbach's character as it deserves.

This book is the product of an almost unique wealth of events and experiences which befell the author at an age which is ordinarily regarded as too young for the moral evaluation of such impressions. Yet there is a surprising maturity in his work and personality. Hence the subjective character of the book, particularly in the extracts from his diary, which strikes readers very forcibly.

But his reflections are more than mere memories. Their documentary value is as great as the author's insistence on objectivity and truth. The book is not only the most comprehensive specialist work on the German Luftwaffe to date, but a convincing refutation of all the legends about sabotage or "a-stab-in-the-back".

Baumbach is a half-way house to subsequent historical investigation and judgement, and his book serves, as Friedrich Meineke once said, as a preliminary to future attempts to understand our fate. In getting together the rich store of documentary material the author had the help of the well-known historian, Professor Bruce C. Hopper, at whose suggestion this book was written.

It provides no sort of encouragement to nationalistic or militant tendencies. The whole story of the tragedy of a service, and the diary extracts, reports and letters, will cure even the most adventure-loving young minds of the idea of "blithe and jolly" wars. The shades of the past which made the author tell a reporter from the publication *Quick* that he loathed war and would never drop a bomb again

[1] Nazi Party. (Tr.)

9

speak too eloquent a language. The book is also a more effective exposure of dictatorship than a hundred well-meant speeches from modern democrats.

Every era is a puzzle which only the future can solve. National Socialism was "the St. Vitus Dance of the Twentieth Century" (Rauschning). It denied individual freedom of conscience. But Man is its representative, not its product. He possesses freedom of decision—that is his eternal right as long as he lives. But he is also faced with the necessity of making decisions—and that is the price he has to pay for his freedom. In a few pages Baumbach draws the conclusions for an uncertain future. He pleads and warns that in the age of the technician "too late?" must not become a real "too late!"

THE AUTHOR

Werner Baumbach, one of the most striking figures in any Air Force in the Second World War, was thirty-four years old when this book was written. At the end of the war he had attained the rank of colonel—and held the post of General of the Bombers. He was born on the 27th of December, 1916, in the little town of Cloppenburg in Oldenburg. He came to the Luftwaffe via gliding.

Scapa Flow, Firth of Forth, Narvik, Dunkirk are the first steps in his unexampled career as a dive-bomber pilot. After a lengthy period of service in the east he was employed as commander of the bomber fleet in northern Norway, scene of the attacks on the Arctic convoys, and subsequently in the Black Sea and the Mediterranean.

Through his friendship with Jeschonnek, the Chief of Staff, and Udet, the Quartermaster-General (Air), he, with a number of other junior front-line officers, was able to bring about a reorganization of the bomber arm. For his services in action he was awarded—the first to be so honoured—the Oak Leaves with Swords to the Knights Cross, the highest distinction to be given to a bomber pilot in the Second World War. He was subsequently commissioned to test new weapons, such as guided bombs. In that capacity he was in almost daily contact with the men at the top. He became a close personal friend of Speer, the Minister of War Production.

In the last phase of the war, in conjunction with Speer, he was able to avert appalling destruction in Germany by his skill in argument, his personal integrity and courage in conferences with Himmler, Göring and Goebbels. At the Nüremberg trials Speer said when giving evidence, "Baumbach, Colonel Knemeyer and I were able to make certain that the latest technical developments in air warfare were brought to the West and their exploitation by the Soviets prevented."

The capitulation found Baumbach in Flensburg-Mürwik. In August, 1945, he was brought to England. He spent nearly six months in an English interrogation camp. He was told that he would be charged as a "war criminal" on the ground that he had fired on shipwrecked people and had been the commander of No. 200 Bomber Group. After unending cross-examination and investigation Baumbach was able to prove conclusively that throughout the war neither he nor any unit under his command had committed any violation of the Hague Convention.

In February, 1946, after further inquiry by American Headquarters, he was released. Professor Dr. Bruce C. Hopper, the Harvard University historian, asked him to assist him in his work. For a whole year they laboured together on studies on the course of the Second World War. Then Hopper suggested to Baumbach that he should write this book.

It was thus that this airman, who since schooldays had had a passion for history and writing, became an author. He was helped by the fact that he had performed the deeds of which he writes so graphically at an age at which "a young company officer hardly dared open his mouth at mess", as Bernard Shaw once put it.

In the spring of 1948, with Allied permission, he emigrated with his wife and son to South America and became technical adviser to industrial firms.

His many-sided activities during the war have given rise to many legends, rumours and conjectures about him, particularly after the war ended. The English Press called him "the German Lawrence of the Second World War". The only true element in them all is that in all his work and actions Baumbach regarded the human side as the only one that mattered, and both during and after the war spoke his mind without regard to any consequences to himself.

When a German reporter called on him after the war he remarked, "I am still an enthusiastic flyer, but only for pleasure. I loathe war and will never drop a bomb again. My military ambitions are a thing of the past."

His ideas about the future development of armaments and war in the air made him one of the accepted international experts on air strategy.

PREFACE

Once again I am glancing through my flying log-book to find a thought fitted to serve as an introduction to this book.

I have written about the war. The facts line up, naked, remorseless. The technical side often seems to thrust the human aspect of the great catastrophe into the background. It soberly and fittingly prefers to leave to the expert and the airman the proper conclusions to be drawn from the questions and problems raised in our exclusive domain. Yet this book has been written not for experts only, but a wider circle whose interest in the nature and development of aviation calls for a popular treatment of a highly complex matter. This compels me to make known my own standpoint as a human being.

In this ticklish undertaking I should not wish to weary the reader with my own short, if eventful, career. Some notes in the last pages of my war diary may enlighten him as to my mental outlook:

"Surely the earth is a Whole only for one who is himself a Whole; it is disrupted and dismembered only to those who are themselves disrupted and dismembered.

"The modern seven-league boots of my beloved bird bore me over blood and mud, pettiness and infamy, from fathomless depths to wondrous heights. Amidst lacy cloudbanks, the fierce storms of the North Sea, the scented dreams of Sicilian nights, the all-knowing smile of the moon, the starry spheres, I began to learn what God is.

"In this twentieth century Dance of Death the door to damnation, the collapse of Western culture, has been forced wide open. Life remains only in the life within, the timid beating of the heart and the painful sting of the heart's longing. And yet Mother Earth will not be wrenched from her moorings nor her framework burst apart: for God still lives and His fires still burn.

"In these years of decision the young airman that I was has greatly changed, and matured far beyond his years. A necessary preliminary was his whole youth, his restlessness, his wanderings, his eagerness for the laurels of heroism, his consuming love, his struggle with his own soul, his travail and despair, his yearning for harmony and salvation."

In this book I am trying to offer a critical commentary on the war in the air, seen within the framework of the war as a whole, with the object of assisting in a subsequent historical examination of the catastrophe to my nation by a truthful relation of the facts. By the title of the last chapter "Too Late?" I do not mean to convey a *post hoc* glorification or a belated dirge for the past, much less a longing for the return of a vanished epoch. Since the war ended these two little words have become catchwords in Germany and the whole Western world. Many "ifs" and "buts" and the most improbable rumours and conjectures—malicious and otherwise—are involved. So far few satisfactory answers as to future development can be made.

As an airman I have endeavoured to see things, and write about them, from a bird's-eye view—not from the public platform. I must take the risk of being misunderstood by many. It is the airman's lot.

The short time that has elapsed since the events in which I participated directly makes a purely objective description scarcely possible and so I make no claim that my conclusions are complete or of universal application. I have written about the war in the air as I knew and experienced it, hoping thereby that in one of the most critical phases of our history I am making a personal contribution to that problem of air power which may well decide the future.

The World War of 1939-45 may well prove to be the last classical war. By all its participants it was fought according to the well-known rules of mass warfare and decided by the rule of superiority of numbers—certainly not by a mere demonstration of what the atom bomb could do. Against the weakness, inherent in the totalitarian system, of German war leadership may be set many shortcomings and failures on the Allied side, which I can do no more than mention in this book. That these did not prove fatal is to be attributed less to the personal qualifications of the victorious Allied leaders than to the greater possibilities open to them.

The German soldier on all fronts, individual military leaders, the German nation as a whole, have performed feats in this war to which no realistic history will ever fail to do justice. The German defeat is only a phase in the collapse of a dying epoch and it should not lead the individual German to despair of the nation's future.

As a specific technical novelty, air power has disrupted the existing political picture, as well as strategic ideas—both for land and sea—associated with it. Air power will win the natural place due to it; in the political and sociological world order it will burst the bounds and bring about general chaos. As air power already meant a change of direction in the conduct of war as a whole— Germany never had a chance to reach the quantitative potential of her enemies but only the possibility of attaining a qualitative advantage which could have decided the war—so it will necessarily bring about far-reaching decentralization in our way of life and therewith the "re-individualization" of man himself.

Modern technique, with the help of which man can be enslaved by man even today, has obviously reached a turning point. While one section of the powers is still trying to slow down the revolving wheel of history while the other is set on making it turn faster, both are ignoring the natural laws of organic growth within the cosmic order. Both efforts, foolish as they are, have led to the vicious circle of successive wars and economic crises which distinguishes this century. The moral decay is its bitterest fruit.

So it seems senseless to measure and judge the responsibility of the individual or even any particular nation by the way in which the world has developed as a whole. It would be much better to find a way out of the dilemma which all men and all nations can take. A new war between the two antagonistic *blocs* will not only tear the defenceless no-man's-land, Germany, apart but mean unprecedented suffering and desolation to all the rest of the world. It would mean no heroic epoch but unimaginable torture for all humanity.

If the world should turn to better ways in the last hour and apply itself to a peaceful New Order and its problems it would have to revise its attitude to Germany. The German nation, tried in many fires and purged of its dross but with its essential substance undamaged, would not refuse to play its part in the world of tomorrow. Air travel could provide a link.

It is from that angle that this little contribution to the history of the war has been written. It is about the war, but not for its own sake but with a view to directing public attention to existing and future dangers. I have deliberately refrained from adding to the number of so-called 'revelatory' publications. All the quotations from anything said or written by leading German personalities, military and otherwise, as well as the principal ideas current in the course of this dramatic story, are available to the public. My own comments are kept in the background lest I hamper the critical judgement of the reader.

PUBLISHER'S NOTE

The abbreviations for the designation of German aircraft are those used by the author. These have been widely used in English, though in aeronautical circles the 109 and 110 are given the prefix Bf instead of the better known Me.

CHAPTER I

ORIGIN AND DEVELOPMENT OF THE LUFTWAFFE

DURING the First World War the idea of an independent air force, corresponding to the army and navy, was unknown. Aircraft were originally employed solely as scouts—the eyes of tactics and strategy. At the outset they were unarmed and not constructed for fighting purposes. It was only in the course of the war that small fighter and bomber squadrons developed.

Air war, the struggle for the military control of three-dimensional space, formed no part of strategy, as conceived by leading soldiers on either side. It was the technical developments during the war that gave the airmen a certain importance, though they were never considered on the same plane as their brothers in the other services. They equipped the aircraft allotted to them for special tasks with technical makeshifts and devices provided by the army—a quite frequent occurrence in the Second World War also. The observers in the first air duels just blazed away at each other with revolvers and rifles, but the technicians soon evolved more effective methods which culminated in shooting synchronized fixed machine-guns through the airscrew blades. "Dog-fighting", i.e. fighter v. fighter, was considered the high spot of air war.

The tactical and strategic use of aircraft remained in its infancy, so much so that the air was never made a theatre of war on its own. Airships and aircraft remained modern adjuncts of land and sea warfare, just like tanks, machine-guns, gas and submarines. The revolutionary idea of a war in which the air arm could play as decisive a rôle as the army and navy had not yet penetrated.

The terms of the Treaty of Versailles involved the dissolution and destruction of the German air force and in accordance with those terms Germany was left with only 140 aircraft and 169 aircraft engines for commercial use only. As the victorious powers also prohibited all aircraft construction, the industry was paralysed and at first Germany was excluded from commercial aviation. The industry had to turn to other activities.

That being the situation, as early as 1920 ex-airmen and eager youth began to devote themselves to gliding. Gliding clubs, private individuals and university groups constructed their own gliders and

2

improved the types from year to year. In this sphere outstanding performances were achieved, performances which were the starting point of the remarkable qualities which subsequently distinguished the civil and military aircraft of Germany and made it possible to maintain our lead over the aircraft of other countries despite all difficulties and limitations.

When the Entente raised the ban on the construction of civil aircraft on the 3rd May, 1922, their restrictions on size and performance remained and the Inter-Allied Air Commission saw to it that they were observed. The German motor industry was unable to acquire experience with heavy aircraft engines as the power and performance of engines in pleasure and commercial aircraft were subject to strict limitations. German engineers were thus reduced to the necessity of studying foreign technical journals and visiting factories abroad if they wanted to get an insight into the developments in other countries.

From 1921 onwards the Reich Defence Ministry took up the question how Germany's military position could be improved, at any rate within modest limits.

It was the Treaty of Rapallo in 1922 which made developments in the sphere of military aviation possible. It contained a secret military clause under which Russia was to be the beneficiary of such developments. The Soviet Union was to make the flying grounds and other requirements available and Germany was to report the progress of the industry and all future developments. In that year the Junkers company became interested in a factory near Moscow and was given a subsidy of eighty million reichsmarks by the German Government. The Heinkel and Dornier concerns also established themselves abroad. Following on the ever closer association between the German Reichswehr Ministry and the Red Army, a training group from the Reichswehr to which specialists in air tactics, bombing, armament and air photography were attached, were sent to the airfield at Lipesk, not far from Moscow. It was there that in 1926 a start was made with the training of German fighter and reconnaissance pilots. The young Soviet General Staff was trained by General Staff Officers of the Defence Ministry. The training and experimental centre at Lipesk (and auxiliary airfields in the Caucasus later on) was maintained until 1933 when Hitler came to power.

The Locarno Treaty of 1926 seemed to usher in a period of relaxed tension in foreign relations. In May there was the Paris Air Agreement. The restrictions on the construction of civil aircraft were withdrawn, though Germany had to make some important concessions in return. The Government had to undertake to spend very little public money on civil aviation. The ban on any kind of military aircraft remained.

The development and testing of aircraft became the province of the civil authorities. Novel aircraft such as the Dornier Do X flying-boat and the four-engined Junkers Ju 38 offered new ideas to international aviation and can be regarded as amongst the fore-runners of monoplanes with four or more engines.

Despite all the energy and ingenuity employed and the un-deniable successes of German aviation in those years, it always had to cope with the greatest difficulties. Every mark meant a fight with the government authorities and aviation invariably succumbed in party squabbles and the personal feuds of leading men. As civil aviation was also denied the valuable suggestions and impetus which military aviation could have supplied, the difficulties were enormously increased.

The Defence Ministry tried to keep itself abreast of the times, at any rate in the field of theory. In a special section of the Ministry ex-army flyers were engaged in the study of foreign air forces. As the small Reichswehr, restricted to a hundred thousand men, could only plan for defence, that consideration necessarily affected the creation of a future Luftwaffe. It was intended to incorporate it in the army. The function of the army air formations would mainly be confined to supporting the ground forces in the concentration and battle areas, so it is easy to see why aircraft suitable for that purpose only were asked for and developed. The German nation and the world had not forgotten the names of Immelmann, Boelcke, Richthofen and Udet, outstanding airmen who in the First World War had laid the foundation stone of Germany's greatness in the air.

As early as 1933, a year in which aviation made great strides, it was clear that to secure mastery of the air a comprehensive know-ledge of flying technique and the art of controlling aircraft both from the air and the ground would be indispensable. In air warfare military planning would have to keep in mind the stocks of equip-ment and its inevitable development as well as the time factor which would become increasingly important having regard to the dynamic character of aviation. Even in the years immediately following the war a few stray voices had been raised to prophesy that a future war would be a total war, but even some of these failed to draw the logical conclusion from that war—that in future science must be given quite as important a place as tactics and strategy. The Italian General Douhet and the French naval engineer Rougeron were the most outstanding advocates of a new theory of air warfare. Their revolutionary ideas were to influence not only the creation of the future German Luftwaffe but the military thinking of all the powers fighting for the mastery of the air in the Second World War.

When the building up of a new German air force was taken in hand, immediately after Hitler came to power, there was no option at first but to draw on the existing types of civil aircraft. There was a repetition of what had happened during the first war, and to some extent the job was put into the hands of the very same men who had carried out the first "adaptations". To fit out civil aircraft for military purposes observers' nacelles were attached like swallows' nests, bomb bays built in, bomb release gear mounted below the fuselage, bomb-sighting apparatus installed, and machine-guns fixed to fire through windows. Thus were the first "multi-purpose" aircraft produced. The German standard commercial plane, the Junkers Ju 52/3m, thus became the first three-engined bomber of the new Luftwaffe.

The question of personnel for the new Luftwaffe was more difficult than that of *matériel*. Captain Hermann Göring, one of Hitler's earliest followers, became Minister of Civil Aviation in 1933 and was simultaneously appointed Commander-in-Chief of the new Luftwaffe. In this capacity he was soon given the rank of general. A number of his former comrades in the war followed him, mounting the promotion ladder at incredible speed, yet there were nothing like enough of the surviving "old eagles" to supply the new Luftwaffe with the high command and staff corps it required. Army and naval officers had to be called on, though the great majority of them had had no practical experience of flying. As the army and navy were themselves in the process of resurrection they became increasingly reluctant and unable to lend personnel to the Luftwaffe. Many officers of the old imperial army and navy were brought out of retirement to fill the most pressing gaps. One result of all these measures was that the majority of the officers in the higher branches of the Luftwaffe were men who did not fly, and most of them naturally were not qualified to supervise and carry out a technical reformation on modern lines. With few exceptions, even those who had flown could not forget their war experiences or abandon their outmoded ideas of air warfare. The non-flyers were better equipped intellectually but were also zealous advocates of the idea of strategic and tactical air fighting, i.e. the employment of aircraft in direct or indirect support of the ground forces. This military thinking was subsequently to square with that of the political leaders.

The High Command was extremely impressed when the range of bombers was extended to a few hundred miles. To army thinking that was an extraordinary achievement, even though the types concerned, the Dornier Do 17M and Heinkel He 111F medium bombers, were barely adequate for the direct support of motorized units.

The fact that Germany subsequently entered the war with the most modern and efficient air force in the world is to be ascribed less to the strategic foresight of the men at the head of the armed forces than the outstanding achievements of her scientists, inventors and technicians. Although the latter were subject to severe limitations, especially as regards aircraft engines, in the years before 1933, and had also to cope with continuous financial difficulties and an almost chronic shortage of raw materials, they achieved revolutionary advances and pioneered developments such as the Junkers wing, the flap, the tailless glider and aeroplane, which reached their apogee in the supersonic projects. Germany's activities in civil aeronautics showed the way to the technical conquest of the air.

In March, 1935, when the German Luftwaffe came out into the open as an independent arm of the new Wehrmacht (Armed Forces) and was raised by Göring to the status of a military-political weapon, the engineers and technicians of our aviation industry succeeded in an astoundingly short time in perfecting the fuselages, engines, weapons and bombs. But as no precise long-range goal was set by the High Command—an absolute necessity for a modern air force designed as the decisive weapon in war—they had to be content with the production of certain prototypes.

One of the few men who realized from the start that in any future war technology would play the vital part was General Wever who unfortunately lost his life in a tragic crash in the summer of 1936 when he was the first Luftwaffe Chief of Staff. "Perhaps in no other arm of the service is the mutual dependence of tactics and technology and their interworking so great as in the Luftwaffe," was one of his maxims which he drilled into the new service at the very beginning.

It was during his term of office, when long-range bombers made their appearance, that we had serious indications of the intention to build up a strategic bomber fleet as the backbone of the German air force and the prerequisite of strategic air warfare. This meant the practical realization of Douhet's idea of reducing the enemy's resistance and weakening his morale by air attack on industrial installations and communications. Wever shared Douhet's conviction that a modern war could no longer be localized or limited to order.

In Wever's time the Dornier concern had developed a four-engined long-range bomber, the Do 19, and in 1936 produced three prototypes. Simultaneously Junkers had available two aircraft of the four-engined Ju 89 type. After Wever's death these experimental aircraft were broken up on the orders of the German Air Ministry and a promising line of research was thereby abandoned. The

Deutsche Lufthansa, the State airline, however, was busy with the development of a four-engined civil aircraft, the Focke Wulf Fw 200 Condor, a type which did good military service in the Atlantic at the beginning of the war, despite its characteristically civil features. It was only under the compulsion of war that in the years 1940 to 1942 the construction of big aircraft was resumed, mainly by Junkers. Work on the four-engined Ju 90, and later the Ju 290, which had been interrupted in 1936 was taken up again and the Ju 252 was put in hand also. But these types were not yet ready for strategical employment as long-range bombers in Wever's sense, for they were primarily large transports.

General Wever was considered as the best brain in the German Luftwaffe. After his death the General Staff had no one else in a responsible position who so clearly realized that air weapons and air strategy were interlocked and took that fact into consideration in his decisions.

After a short interregnum under Generals Kesselring and Stumpf, a successor was appointed in the person of Hans Jeschonnek, subsequently Colonel-General Jeschonnek. He was a convinced advocate of the view that light and medium bombers should provide the core of the Luftwaffe. In this he was in agreement with the head of the Luftwaffe, Göring, the O.C. Air Armaments, Lieutenant-Colonel Udet, and the majority of their advisers in the Technical Branch, Air Ministry. Jeschonnek's views were determined by his knowledge of Germany's limited raw material and production capacity, particularly in respect of aluminium, engines and instruments. The ever-strained fuel position of the Reich also strengthened his opinion that Germany could not produce the large number of multi-engined aircraft which would have been required for "carpet" bombing with the technical resources then available. He believed that precision bombing in level flight or, more particularly, diving, carried out by crack crews in crack aircraft, was more suited to the Luftwaffe. In this he received strong support from Udet who had practised dive-bombing in Germany with two aircraft bought in the United States and must be regarded as the pioneer of that art. So immediately after Jeschonnek took over, the General Staff issued orders to industry that all aircraft must be able to dive.

The factor of limited time was not the only difficulty which the Luftwaffe had to cope with. Its organizational reconstruction and the development of its aircraft and armament were also hampered and delayed by the stream of ever changing and frequently conflicting requirements of the General Staff. The one consistent strand in the whole process was the emphasis placed on bombers, i.e. an *offensive* weapon.

The resurrection of the Luftwaffe was certainly an impressive spectacle for onlookers. Airfields and barracks were built with breath-taking speed. New airframe and engine factories were literally stamped out of the ground. In 1935 the S.A. gave Hitler as a birthday present a complete fighter wing which took the name of the "Horst Wessel Wing", to symbolize the spirit of the new age within the Luftwaffe as well. In less than a year we had air war schools and an air war college for the training of an air staff. In Berlin the giant buildings of the Air Ministry literally rose out of the ground. On Hitler's next birthday after the official start Luftwaffe squadrons flew low over the buildings lining the east–west axis, demonstrating the menace of Germany's might to the foreign visitors present.

Within a very short time the determination of Hitler and Göring made the country an air power, the anticipated effect of which was not lost on the rest of the world. But what did the men at the top think of the efficacy and possibilities of the new arm? In the spring of 1939 Jeschonnek delivered himself as follows at the conclusion of a staff course, which had a campaign against Poland as one of its subjects:

"In the Luftwaffe it is not just a question of technology. In the technical field every state is always trying to get ahead of every other. But we must realize that all states are really on the same technical level and that there is no such thing as a permanent lead. But the development of air tactics is so recent that in that field conclusions can be reached which, translated into action, could mean actual superiority over the enemy. The duty of the General Staff is to indicate to the technicians the requirements they must meet, but its most important task is to make the best possible use of what the technicians give them, to extract the maximum out of men and machines at the lowest possible cost.

"The further development of air power must take economy as its motto, economy with *matériel* even more than with money

"I will end by asking all General Staff officers to do their best to keep in touch with practical and technical developments, despite all the claims on your time and attention. This course will have fulfilled its purpose if you appreciate that red-tape must be avoided and the vital thing is to work out how you can reach your goal at the lowest possible cost."

Jeschonnek never fully realized how important it was that the Chief of the General Staff should have his say in the build-up of the air fleet. The real cause of the progressive aimlessness in the course of the war, with the resultant growing friction between the different departments was, the absence of comprehensive planning which entailed rigid adhesion to decisions once taken. The development of air-force equipment and armament would doubtless have fared better if the General Staff had kept in the closest touch with the technical department. What the German Luftwaffe and its reconstruction lacked was the co-ordinating expert—the military-cum-scientific head.

The German political and military leaders were convinced that only well armed forces, and in particular a strong air force, could ensure the safety of the country as things then stood. The view—today so hotly disputed—that power, especially air power, is the surest guarantee of peace, was even then being put forward in Germany. In the years 1930 and 1931 it was estimated that the nations of the world had spent four to five thousand billion dollars on "the Moloch of Rearmament".

The rearmament avalanche was in progress before Hitler took over the reins, so the first and sole preoccupation of Germany's new rulers was to catch up with the armaments of the rest of the world and, if possible, surpass them. They had only to follow the example of their rivals. The undertaking had of course more serious economic consequences for her than for the much richer great powers. Göring's often-quoted remark "guns before butter" was particularly applicable to this phase of our rearmament, which even before the war led to a marked shortage of consumer goods of all kinds in Germany. By 1939 the avalanche of world rearmament had reached its climax and the rearmers had obviously lost control.

When the wiseheads asked where all this arms competition was leading Göring answered them in 1935: "The engine drone of German fighter and defence squadrons will not disturb the symphony of peace. Quite the contrary; it will be no more than a new harmony in the honourable *leitmotiv*, 'Germany means peace!'" Two years later he repeated the same theme: "Germany has made her contribution to the peace of the world. Germany's main contribution to world peace has been her rearmament."

In 1938, at the time of the Sudeten crisis, he warned the unwary: "The Luftwaffe will be one of the guarantees of peace for Germany, but I frankly admit that it will be a terrible day if it gets the order to strike. On that day we will swear to the nation that we will be the terror of the aggressors and nothing shall hold us back from neck-or-nothing."

On the 1st March, 1939, Göring made a confession: "Since the 1st March, 1935, I and my colleagues, carrying out the Führer's intentions, had created at high speed the most modern air force which any nation could possess. I am proud that the German Luftwaffe can serve as a powerful instrument of the Führer's creative statesmanship. We shall know how to maintain the advantage. I am for ever convinced that we cannot be conquered by force of arms. Fear of our invincible air squadrons and our ultra-modern, splendidly trained flak[1] artillery has given many a hate-filled warmonger abroad bad dreams.

"Our worried enemies found themselves faced with the fact that Germany possessed the mightiest air force in the world. There can be no doubt that this fear has helped to restrain such warmongers from war; they could not block the peace-loving statesmen's road to our Führer and a fair understanding.

"And then came the days of crisis, the decision for war or peace. And the old truth was confirmed once more—peace is preserved only by a sharp sword. Without a sharp sword peace is utterly fragile. They know abroad that we were not bluffing."

And yet how often we bluffed! When Galland and I were detained in England after the war he once had this to say about the reoccupation of the Rhineland: "It was a masterpiece of Hitlerian bluff. At that time all we had to fly with in western Germany was one Arado Ar 68 biplane fighter squadron (incidentally it was unarmed), in which we flew from aerodrome to aerodrome. At each stop the insignia of the aircraft were changed so that a 'new squadron' appeared at the next aerodrome and the new German Luftwaffe could be displayed to the assembled local press."

As the war went on Göring and his immediate associates ultimately became victims of their own bluff that "the Luftwaffe's advantage was assured for all time", and that "its start could never be overtaken, whatever might happen. . ." It was Göring himself who never freed himself from the memories of his feats as a fighter pilot in the First World War, feats for which he won the *Pour le Mérite*. He always had the greatest difficulty in evaluating technical innovations. To him the achievements of a scientific invention were interesting toys or imposing instruments of power over which he could wax enthusiastic. A right decision or critical judgement in technical matters was often a matter of pure luck in his case. He was also at the mercy of his own moods, a fact which contributed to the great lack of overall planning which distinguished the development of the German Luftwaffe and its armament before and during the war.

At that time the views of Ernst Udet, the chief of the Technical Department' were as follows:

"With regard to the basic principles for the production of the most efficient materials, first and foremost is the age-old presumption that the rate of technical progress in weapons is of vital importance to the current superiority or inferiority of a state *vis-à-vis* its rivals.

[1] Anti-aircraft (Tr.), a word later to become common British usage.

From the military point of view, that state has superiority which is in a position to be first in incorporating the latest achievements of scientific and technical progress in its armaments. But for various reasons it is not possible to use all the technical discoveries in the production of new aircraft straight away. In the first place, the ordinary exigencies of air activity prevent immediate attention to every new technical idea, and there are also considerations of economy and supply which impose a wise restraint."

Udet, though a gifted flyer and champion of the view that no one who does not fly himself is in a position to form sound judgements in such a matter, was not the man to tear apart the spider's web in the Air Ministry. He knew that in the sphere of technical progress a man's own feelings, his creative force, his own convictions and love of responsibility, played the dominant part. But it was not in him, in face of the eternal dissension in the departments of that Ministry, to force his ideas on the "Engineer-Generals", fighting and intriguing for power, and our industry and economy. It irked him to be spending his time at a desk instead of in the air. An artist by temperament and inclined to blind confidence in others, he left many decisions to his colleagues, trusting implicitly to their decency and good sense, capacity and knowledge. The Technical Department soon became a department in which the "Engineer-Generals" did pretty much what they liked. At a time when revolutionary discoveries, especially with regard to armament and the effect of jet and rocket propulsion on speed, had been made and were clamouring for materialization, the personal vanity of the leading men in the Luftwaffe made them eye each other with suspicion and quarrel over silly questions of authority. Even before the outbreak of war there were some unhappy revelations in the Luftwaffe, especially at the top. Udet ultimately went to pieces over the inner conflict between his soldierly duty and his conscience; being in a false position, he became unable to cope with intrigue and fateful decisions.

The death of Colonel-General Udet, a man who was held in high honour by us juniors, was a great blow to me personally as he had honoured me with his friendship. He was one of the really sincere men who had warmed up the cold stone building of the Air Ministry, with its thousand main rooms and as many anti-rooms.

I quote from my diary: "Now Ernst Udet is dead. A few days before All Souls' Day he shot himself in his house. I am to attend the state funeral. As I threw away my top boots on my flight over a Finnish swamp I cannot be one of the actual bearers and don't need to borrow a steel helmet. But I am to sit next to the relatives in the solemn setting of the Great Hall of the Air Ministry...

"The hall was filled with the scent of fresh flowers, as always on the days of christenings or burials. The higher staff of the Luftwaffe and representatives of the Party, State and Services were grouped in the background in order of rank and seniority. Jeschonnek looked across to me and gave me a significant nod. Oshima, the Japanese Ambassador, appeared in the doorway and bowed deeply to the coffin. A brief, soldierly gesture. Immediately behind him was Goebbels, in a dark brown leather cloak too long for him and holding his cap in his left hand—not by the peak, as a soldier would, but by the lining. He raised his right hand in greeting to the assembly and stood stock-still for ten seconds. The Führer's wreath was brought in and placed on the bier so that everyone could read the words 'Adolf Hitler'. The last to appear was the Reichsmarschall [Göring]. Red-brown boots, light grey uniform and smart gold braid. He delivered the funeral oration. The bearers, drawn only from officers holding the highest distinctions, stood at attention with drawn swords round the bier, their eyes fixed rigidly on the white plastered wall opposite. The only things that moved were Galland's moustache and a standard. Galland seemed to be missing his cigars. The standard-bearer was the worse for standing too long and had to be changed.

"Göring stepped forward. He looked round, called up his aide-de-camp, gave him a nasty look as he asked him a question and angrily shook his head. But it was too late for orders. From the high balcony behind the assembled company the 'little Beethovens' struck up the phrase from the *Eroica* selected for the occasion. Then Hermann Göring, Reichsmarschall of the Greater German Reich and Commander-in-Chief of the Luftwaffe, ascended the dais. The spurs—too big—clanked and the press cameras whirred. The Reichsmarschall raised his voice. In moving words he spoke of Ernst Udet, the flyer hero. His audience was obviously deeply impressed.

"My thoughts took wing. Who really is a hero? I wondered. Around me are gathered all the air aces of this war. I am told that I am the most successful fighter of them all. Those words will appear tomorrow on the front page of the *Berliner Illustrierte*. It may even be true. But I no longer know what is true and what is not. I looked into the faces of the others—Lützow, Oesau, Pelz, Galland and a dozen more. We are not all heroes. Good shooters and bomb droppers perhaps. But heroes exist only in legend.

"The Reichsmarschall spoke of his only friend, who had been his finest comrade in the First World War, and also of the irreplaceable loss the Luftwaffe had suffered. In the middle of his speech he broke down with emotion over the words: 'I can only say I have lost my best friend.' A *tour de force* on the part of actor Hermann Göring."

There were many stories about Udet and the manner of his death. Only one of them is true. He took his own life of his own free will and was not the victim of a fatal accident while testing a new weapon. His death was due to serious friction with Göring, difficulties with his opponent, Milch (who succeeded to his post a few days after his death), and hopeless complications with the aircraft industry, and especially his former friend Messerschmitt.

A letter which Udet wrote a few months before his death throws light on the human tragedy of this great-hearted airman and personality. On the 25th July, 1941, he wrote:

"Dear Messerschmitt,

"Though I fully realize the achievements of your aircraft, which are of outstanding importance in the battle line itself, I feel impelled to impress upon you that in my opinion you are going ahead on wrong lines. With military aircraft, one must—particularly in war times—design on safe and solid lines and not be compelled to resort to subsequent time-wasting expedients. If that is done, relations between development and serial production will also be more harmonious.

"Not only as Quartermaster-General of the Luftwaffe—responsible for the due and punctual appearance of new models as well as for their performance—but more particularly as your friend, I feel it my duty to point out that the path you have chosen is dangerous and can involve us all in the greatest difficulties.

"It is just the preferential position you enjoy as an aircraft designer, which has hitherto meant unlimited confidence in you and brought you high honours, which should intensify your sense of responsibility and urge you to look upon yourself and your achievements with a more critical eye.

"This month once again we have received no Me 210s for battle tests, so that the employment of the model has to be postponed for another month.

"I also have the impression that since the first prototype was built you have made far too many changes. The prototype and the proposed machines in the series are so different that the results of testing are of no use for the practical question of behaviour in action.

"One thing, dear Messerschmitt, must be made quite clear between us, and that is that there must be no more losses of machines in normal ground landings as the result of a faulty undercarriage; this can hardly be described as a technical novelty in aircraft construction.

"All these unnecessary scandals and this waste of time call for higher standards in the testing of your new aircraft, and I shall report in that sense to my department.

(Signed) Udet."

There were a few more letters in which Udet tried again and again to preserve his friendship with Messerschmitt. The dead cannot speak. Their judge is not of this world. But the living must justify themselves. Udet's death shook Göring's prestige with Hitler. Though the real background was certainly kept from the Führer he had far too great a knowledge of men not to suspect what had happened or to realize who were the guilty men. So the latter, Göring in particular, lost no time in superfluously justifying themselves in public and ascribing the failure of the Luftwaffe to Udet.

"If only I could find any explanation of what Udet was really about," said Göring on the 9th October, 1943. "He brought our air service into complete chaos. If he were alive today I should have to tell him that he was the destroyer of the Luftwaffe."

During the failures from 1943 onwards Field-Marshal Milch too often excused himself by reference to the shortcomings of Udet. But neither Göring nor Milch had any right to complain as Udet had asked again and again to be given other employment. "Don't worry your head about it," Göring once said to him. "I'm assigning a few good engineers to you and you'll manage all right." Now the child had fallen in the fountain.

Not personal failures but fundamentally unsound decisions, based on ignorance of the true nature of air power, were responsible for the Luftwaffe's inability to cope with its tasks in the subsequent testing period of the war. The improvisations forced on us by the lack of time and the acute shortage of raw materials were by no means a minor factor. In the few years devoted to the build-up of the Luftwaffe more than was humanly possible was done to give the German Reich an Air Force adequate to create respect and secure its position among the nations.

CHAPTER II

PRE-WAR STRENGTH OF THE AIR POWERS

THE German air leaders, Hermann Göring himself most of all, were firmly convinced of the technical and numerical superiority of the German Luftwaffe. They also believed that that superiority over the other air powers could be maintained for years. This over-estimate of our own strength, and under-estimate of that of our prospective enemies, is betrayed not only in the political speeches of leading men in these critical spring months in 1939 but in the archives of the General Staff of the Luftwaffe. Here, for instance, are the comments of Lieutenant-Colonel Beppo Schmitt at the beginning of 1939 on the stage reached in the rearmament programmes of Europe and America:

"English and French air fleets are still much out of date. British air defence is still weak. In 1940 we may expect a monthly output of 300 English and 200 French front-line aircraft. In the next few years it is not to be anticipated that they can catch up with German capacity. The U.S.A. has only weak air force, and no increase possible for at least six months. England will not be able to get out of a fight with the German Luftwaffe. German aircraft are superior in view of their advantage in armament, armoured petrol tanks and flying instruments. In Germany alone has an overall view of war in the air been taken."

In the spring of 1939 the same section of the General Staff issued a comprehensive report on the assumed strength of the European air powers as well as the U.S.A. It contained the following figures:

Great Britain.
  5,500 (Royal Air Force and Royal Navy aircraft) of which 3,600 are front-line aircraft of the independent Royal Air Force (and of which 20% are first class) composed as follows:
    2,500 are bombers (only 500 first-line);
      620 are fighters (200 first-line);
       30 are dive bombers;
      450 long-range reconnaissance planes (90% second-line).

The equipment must be regarded as very largely old-fashioned. It looks as if "highly-primitive ideas" about war in the air hold the field there.
Experience of transoceanic flight must be regarded as inadequate.

Anti-aircraft defence lacks all proper equipment. There are about 600 heavy and 2,800 light guns, and about 3,300 outmoded searchlights.

The Royal Air Force is far behind the German Luftwaffe as regards operational readiness, particularly in air defence.

It seems that the rearmament goal England has set before her cannot be reached in any near future, so orders have been placed in the U.S.A. and aircraft industries established in Canada and Australia.

In the period up to the 1st April, 1940, the R.A.F. is aiming at an output of:

Home—about 2,400 aircraft.

Overseas—about 500.

Naval—about 500.

Anti-aircraft artillery is to be doubled. It remains to be seen whether this is really intended, or whether the statements in Parliament and the Press are not mainly a sop to a highly-agitated public opinion.

France.

A total of 4,650 aircraft (including land and naval planes) including 2,500 front-line aircraft of the independent *Armée de l'Air.* Only about 30% of these can be regarded as first-line. They are divided into:

1,300 bombers and reconnaissance aircraft;

150 ground attack aircraft (all first-line);

1,100 fighters (with 400 first-line).

Both the fighting service and the anti-aircraft artillery are equipped with old-fashioned material. The aircraft industry is still far short of the Government's intention.

Orders have also been placed in the U.S.A. and Holland. By the middle of 1940 it is expected that about 3,800 aircraft, including 700 from abroad, will have been delivered. It is planned to double the anti-aircraft artillery.

Belgium and The Netherlands.

Rearmament must be regarded as inadequate.

U.S.A.

The Air Force is numerically weak, though equipped with up-to-date models.

It must be assumed that in case of a war the U.S.A. will not sell its aircraft at once. But we must presume that if Great Britain and France were hard-pressed, aircraft would be supplied.

Hitherto the aircraft manufacturing companies have been starved.

Spare parts non-existent. Now the industry is being urged to expand by the Government, which has apparently no means of its own of doing so. At the beginning of the war we need not anticipate any effective support from the U.S.A. Six months later it might be different.

Poland.

The Polish Air Force comprises the Army Air Support Air Force and The Tactical Air Force with:

900 aircraft, which consists of about

270 fighters (30 first-line);
170 bombers (130 first-line);
175 reconnaissance (all first-line);
190 army co-operation (all first-line).

The anti-aircraft defence is weak.

Supply arrangements defective.

The aircraft factories are in the vicinity of Warsaw. They cannot increase output.

Russia (Soviet Union).

Information about Russia is very scanty. The strength of the Russian Air Force is about 6,000 aircraft, of which about 5,000 are front-line machines.

In Europe there are about:

1,300 bombers and long-range reconnaissance aircraft (800 first-line);
350 ground attack planes;
1,400 fighters (1,200 first-line);
250 observation aircraft.

The remaining aircraft are stationed in the Far East.

Naval aircraft, about 700, mostly out-of-date, are mainly in the Baltic area.

Anti-aircraft artillery—bad.

The aircraft industry can only meet current requirements.

Denmark, Sweden and the Baltic States.

Of no significance, except that, Sweden and Finland in particular, may acquire a certain importance as suppliers of raw material.

Balkan countries.

Of no importance.

Italy.

About 3,800 aircraft (land and naval planes) of which 2,800 are front-line.

By 1940 the whole Italian air force can be considered out of date.

The anti-aircraft artillery has only out-moded equipment. The aircraft industrial capacity is inadequate.

This report concludes as follows:

"The Western Powers are rearming in order to catch up with Germany in the next year or two. If they did so, the position in the air, particularly as regards air defence, would be changed. But any increase on the present figure schedule can be regarded as impossible.

"The present German qualitative level will be reached with the Spitfire, Hurricane and Morane 406 fighters, as well as heavy anti-aircraft guns. But the bombers remain inferior. American help will arrive too late."

The wording of the General Staff report, from which these extracts come, shows that the German military leaders were accepting the possibility of an armed clash in Europe in the near future.

As regards the rearmament and effective strength of the German Luftwaffe at the outbreak of war, the official report of the Quartermaster-General of the Luftwaffe in September, 1939, gives the following figures:

Thirty Bomber Wings—1,180 medium bombers.
 18 equipped with Heinkel He 111F and He 111P aircraft;
 11 equipped with Dornier Do 17M aircraft;
 1 equipped with Junkers Ju 86G aircraft (obsolete).
Thirteen Day Fighter Wings—771 single-seat fighters.
 12 equipped with Messerschmitt Me 109E aircraft;
 1 equipped with Arado Ar 68 biplanes.
Nine Dive Bomber Wings—336 dive bombers.
 9 equipped with Junkers Ju 87A and Ju 87B aircraft.
Ten Attack Wings—408 "destroyers".
 10 equipped mostly with Messerschmitt Me 110C twin-engined fighters, but some with Messerschmitt Me 109D aircraft.
One Army Support Wing—40 dive bombers.
 1 equipped with Henschel Hs 123B aircraft.
Two Transport Wings—552 transport aircraft.
 2 equipped with Junkers Ju 52/3m aircraft.
 The Lufthansa, training schools, etc., aircraft would also be impressed into service.
Twenty-three Reconnaissance Squadrons—379 reconnaissance aircraft.
 23 equipped for the most part with Dornier Do 17P aircraft.
Thirty Army Reconnaissance Squadrons—342 scout aircraft.
 25 equipped with Henschel Hs 126B parasol monoplanes;
 5 equipped with Heinkel He 45 and Heinkel He 46 aircraft.

Eighteen Naval Squadrons—240 aircraft.

14 Coastal Squadrons, equipped with Dornier Do 18, Heinkel He 115, Blohm und Voss Bv 138, and Arado Ar 196 aircraft

2 Shipborne, and 2 Aircraft Carrier Squadrons, equipped with Heinkel He 59 and He 60 aircraft.

Sundry Units—55 aircraft.

Total number of aircraft available: 4,333 aircraft.

[N.B. German aircraft units were made up differently from those of the Royal Air Force, and can be summarized as follows:

Two aircraft to each Section (*Rotte*), and two Sections to each Flight (*Schwarm*), with three Flights to each Squadron (*Staffel*), and three Squadrons plus a few extra aircraft made up a Wing (*Gruppe*) of about forty aircraft on the average. Three Wings were made up into a Group (*Geschwader*) with 90 to 120 aircraft, depending on the type and rôle. Above this came the Divisions or Corps which comprised differing numbers of Groups depending on rôles; an Air Force Division (*Flieger Division*) or Corps (*Fliegerkorps*) comprised Groups of aircraft of different functions, but a *Jagd Korps* or *Jagd Division* consisted of fighter aircraft only. Divisions or Corps were combined with an Air District (*Luftgau*) which looked after the administration side to become an Air Fleet (*Luftflotte*). The Air Force (*Luftwaffe*) had five main Air Fleets, each under the command of a General, and there were ten major Corps, although not in numerical order or in the same proportion necessarily in each Air Fleet.—Tr.]

Anti-Aircraft artillery:

About 650 heavy batteries (88 mm. and 105 mm.) with about 2,600 guns;

About 560 light batteries (20 mm. and 37 mm. cannon) with about 6,700 guns;

About 188 searchlight batteries, with about 3,000 searchlights.

The German Luftwaffe was armed only for a European war with limited objectives and of short duration, i.e. a "blitzkrieg". At that time the General Staff took the view that Germany, by a lightning attack from the air, was in a position to destroy the enemy air forces on the ground and paralyse their air activities long enough to enable the army to occupy their countries. In the case of England, it was believed that mastery of the air over the British Isles could be secured and any revival of air opposition effectively prevented.

CHAPTER III

HIS IRREVOCABLE DECISION

"I HAVE called you together to give you a picture of the situation, so that you may appreciate the individual elements on which my decision to act is based and fortify your confidence.

"It was clear to me that sooner or later it would come to a showdown with Poland. I had decided to act last spring, but reflected that within a few years I should have to turn against the West and then deal with the East. But one cannot be pinned down to a particular timetable. Nor can one close one's eyes to a threatening situation. I would like to have established tolerable relations with Poland in order to fight the West first. But this plan which I preferred has proved impractical because the position has changed in material respects. It has been clear to me that Poland would attack us while we were fighting it out with the West. Poland seeks an outlet to the sea. The new development became apparent after our occupation of the Memel region and it was clear to me that under certain circumstances a war with Poland might come about at an unfavourable moment. My reasons for this view are as follows:

"First, two personal factors; my own position and then Mussolini's.

"By and large everything depends on me, my continued existence, my political knowledge and abilities. Then there is the fact that no one else possesses the confidence of the German people to the same extent as myself. In the future no other man will ever have as much authority as I have. My existence is thus a vital factor. Yet I can be killed at any time by some criminal or lunatic.

"The second personal element is the Duce. His continued existence is also of decisive importance. If anything happens to him Italy's loyalty to the Axis will become uncertain. The Italian Court is fundamentally hostile to him. Above all, that Court regards the extension of the Empire as a burden. The Duce has the strongest nerves in Italy.

"The third personal element favourable to us is Franco. From Spain we can only claim a benevolent neutrality. But this depends on Franco himself. He guarantees the preservation of the present system in Spain. We must accept the fact that there is no other Fascist party with the internal strength and solidity of ours.

"On the other side the picture is negative so far as the leading

figures are concerned. There are no outstanding personalities in
England and France.

"With us decisions are easily taken. We have nothing to lose and
everything to win. In consequence of our limitations our economic
position is such that we can only hold out for a few years. Göring
will confirm this. We have no other course open but to act. Our
opponents risk much and have little to gain. England's stake in a
war is enormous. Our opponents have leaders who are below
average. No personalities. No masters. No men of action.

"In addition to the personal factors the political situation is
favourable for us. In the Mediterranean rivalry between Italy and
France and England; in the East tension which is frightening the
Mohammedan world. In eastern Asia tension between Japan and
England. The English empire was not strengthened when it
emerged from the last war. Nothing had been gained on the mari-
time front. The Anglo-Irish conflict. The independence of the South
African Union has been strengthened. Concessions had to be made
to India. England is very seriously threatened. Unsound industrialization. A British statesman can only be anxious when he looks into the
future. France's position has also become worse, particularly in the
Mediterranean. . . .

"The founding of Greater Germany was a great achievement
politically; on the military side it was questionable as it was
attained by bluff on the part of the political leaders. It is necessary
to try out our military power. If at all possible, not in a general
showdown but by the solution of individual problems.

"Our relations with Poland have become intolerable. My previous
policy conflicted with the views of the nation. My proposal to
Poland—Danzig and the Corridor—was thwarted by England's
intervention. Poland altered her tone towards us. The tension will be
unbearable in the long run. The law of action must not be left to
others. The moment is more favourable now than in two or three
years. A murderous attack on me or Mussolini could alter the
position to our disadvantage. People cannot sit facing each other
with loaded rifles for ever. A proposed compromise solution is said
to have demanded a change of mind and a kind gesture from us. It
is the old Versailles tone of voice. There is the danger of loss of
prestige. At the moment there is still a great probability that the
West will not interfere. We must take the risk with ruthless determination. We are faced with the hard alternative of striking a blow
ourselves or being certainly destroyed sooner or later. I have always
taken great risks in the conviction that success was possible. It is a
big risk now. Iron nerves, iron resolution.

"The following special reasons confirm me in my view:

"England and France have pledged themselves. Neither is in a position to do so. In England there is no real rearmament but only propaganda. Many Germans who did not agree with me have done a lot of harm after the solution of the Czech question by telling and writing to the English that the Führer had proved right because they lost their nerve and capitulated too soon. It explains the present propaganda war. The English talk about a war of nerves. One element of this war of nerves is a parade of rearming. But what is the actual position of British rearmament? The 1938 naval construction programme is not yet completed—only the calling up of the reserve fleet and the purchase of fishing boats. There will be no material additions to the fleet before 1941 or 1942.

"On land little has been done. England is not in a position to send more than three divisions at most to the Continent. As regards the Air Force something has been done, but it is only a beginning. Anti-aircraft defence is in the preliminary stage. At the moment England has only 150 A.A. guns. The new A.A. gun is on the drawing board and it will be a long time before there is adequate production. There is a shortage of officers. Yet England is vulnerable to air attack. In two or three years the position may change. At the present time the English air force has only 130,000 men, France 72,000, Poland 15,000. England does not want the war for two or three years.

"The following is characteristic of England: Poland wanted a loan from England, for rearmament, but England only gave credits, to make sure that Poland would buy in England, even though England cannot deliver. It indicates that England does not really want to support Poland. She is not risking £8,000,000 in Poland although she has poured five hundred millions into China. England's position in the world is very precarious. She will not take any risks.

"France is short of men. Little has been done in the way of rearming. The artillery is out of date. France did not want to get involved in this adventure. The West has only two possibilities in fighting us:

1. Blockade. It will be ineffective because of our self-sufficiency and additional source of supply in the East.
2. An offensive in the West from the Maginot Line. I consider that impossible.

"There is the possibility that the neutrality of Holland, Belgium and Switzerland might be violated. I have no doubt that all these states and Scandinavia too would defend their neutrality with all their might. England and France will not violate it. So in fact England cannot help Poland. There remains the question of an

attack on Italy. Military intervention is out of the question. No one
contemplates a war of long duration. If Herr von Brauchitsch had
told me that he needed four years to conquer Poland I should have
replied that in that case we would forget about it. We will hold back
the West while we conquer Poland. We must always remember our
vast production achievement. It is much greater than 1914 to 1918.

"Our opponents are still hoping that after the conquest of
Poland Russia will enter the war against us. Our enemies have
reckoned without my great determination. Our opponents are little
worms. I saw them in Munich. I was certain that Stalin would never
accept the English offer. Russia has no interest in the continued
existence of Poland; anyhow he knows that it will be the end of his
régime if his soldiers return from the war either victorious or
defeated. Litvinoff's dismissal was decisive. I have gradually brought
about the change in our attitude to Russia. We begin to talk
politics in connection with the Trade Agreement. Four days ago I
took a particular step which brought it about that Russia answered
yesterday that she was prepared to enter into an agreement. Von
Ribbentrop will conclude it the day after tomorrow. We have now
got Poland where we want her. We need not worry about any
blockade. The East will send us wheat, cattle, coal, lead, zinc. It is
a great achievement which called for much effort. I have only
one worry, and that is that at the last moment some fool will put
forward some plan of compromise.

"Our political aims go further. We have made a start with the
destruction of England's predominance. The way for the soldiers is
open now that I have made the political preparations. Today's
publication of the Non-Aggression Pact with Russia has come like a
thunderbolt. The effects are incalculable. Stalin himself has said
that this course will be of benefit to both countries. The effect on
Poland will be enormous."

In this speech, produced by the United States of America at the
Nüremberg trial in 1946, Hitler informed the heads of the three
sections of the armed services of his irrevocable decision.

Of increasing historical importance is the too frequently disregarded and insufficiently explained fact that it was the attitude
of the Kremlin which determined Hitler's decision, that decision
which was to prove so fateful for Germany and the world. He could
not start a war with prospects of success unless Germany's rear was
covered by a benevolent Soviet Union, as provided in the Molotov–
Ribbentrop Agreement.

For the second time in a generation the twentieth-century sky was
set aflame. There was no lack of inflammatory materials, if only
because the doctrine bequeathed by the statesmen of Versailles

contained the seeds of a future conflict. For the second time the
apocalyptic horsemen were to sweep across the world, bringing
famine, death and ruin, because men who had been struck blind let
failing reason convince them that their vital problems could not be
solved by peaceful means.

It was Hitler's totalitarian régime which for the first time sought
to exploit the triumphs of science to aid government in peace and
bring victory in war. It was based on the following assumptions:

1. The home country would be ready for any sacrifice and any
   effort.
2. The maximum concentration of force, arising from the fact
   that the direction and control would be in one hand.
3. A large number of qualified technicians, at least equal if not
   superior to those of other countries, existed.
4. A labour force was available in which any idea of striking was
   unthinkable.
5. There was the possibility of using practically the whole of
   Europe's industry for our own arms industry.

There can be no doubt that at that time the National-Socialist
Germany was far superior to its Western opponents as regards
concentrated force and energy. Monsieur Reynaud, the last Prime
Minister of the Third French Republic, accurately described
Western Europe's position *vis-à-vis* totalitarian Germany when he
spoke of the democratic powers as "in discord, failing courage and
lack of resolution".

Germany's advantage was unquestionably the work of Hitler,
whose imaginative determination had in a few years mobilized all
the spiritual and material reserves of a beaten nation. Himself a
child of the people, he knew better than anyone else how to awaken
the thoughts and feelings slumbering in the hearts of the masses,
put them into words and make them political power by translating
those words into deeds. Yet his military ideas, in which he often
revealed genius, were land-bound, born of the experience and
memories of the First World War. Just as politically he followed
the maxims of Machiavelli—like all other national states in modern
times—so his strategic ideas were to prove equally Continental.

But the totalitarian system, thanks to the rigid, dogmatic thinking
of its leaders and the clumsiness of its organization, was not equal to
the enormously increased pace of development which the war
brought about.

Even in the previous war it was still possible for a man to plan
battles and campaigns on a little map. It took him even less time to
issue his orders. But even then it was a great gamble and involved an
even greater danger!

Battles and campaigns could still be won with a vast expenditure of materials and human lives. But the decision was no longer brought about exclusively on the fighting fronts but at the strategic rear. It was not alone the strength of the armies or the courage of the soldiers on which everything depended but the available raw materials, the inventiveness and productive capacity of the home industries, and above all the inquiring minds of scientific and technical geniuses. For the success of their creative work the latter needed time, freedom from interference, adequate resources at their disposal and sufficient protection, especially against attack from the air. Little time could be given them because rapidity of action was always imposed on the Reich. Freedom from interference was limited by the totalitarian system. Large resources were apparently available. But protection against attack from the air was much neglected in the course of the war.

By and large, a vast armaments apparatus was created under Hitler, but it was not an organic growth and it was always bound to be inferior to that of the U.S.A. unless priority was given to the new inventions. In fact, such priority was either not given at all or given too late.

In addition to the unfulfilled requirements in Germany there was another factor, the incalculable factor of luck, which was necessary for a victorious conclusion to the war. As one of the most striking examples of luck in the Second World War the case of the Allied bomber fleets may be cited. The American four-engined B17 Flying Fortress was by no means the last word in aircraft construction. Judged by ordinary standards of aeronautic development it was long out-of-date when it appeared over Germany. Even the American technique of daylight mass attack at relatively low altitudes and speeds must be regarded as old-fashioned.

Despite the handicaps and limitations in Germany before and during the war to which I have referred, revolutionary technical ideas were developed which would have enabled us to put a speedy end to the four-engined bomber. Yet the enemy succeeded in reducing Germany to dust and ashes, paralysing our war industries, producing a general collapse before the war was over and so attaining final victory. How did this happen?

CHAPTER IV

GERMAN AIR ARMAMENT

EVERY rearmament programme is substantially determined by an interplay of technical factors, politics and strategy. All three factors must be considered and dealt with in concert if the programme is to be successfully realized so that the state will always be in a condition of complete preparedness. The technician has to adapt his planning to the political and military objectives aimed at by the leaders of the state, and the latter in their planning must bear in mind the technical development and productive capacity of their industries, the raw material position in their own country and the possible achievements open to their labour and military forces. All of them, technicians, politicians and soldiers, have to regard time as a material factor in the rivalry with the technical progress of other nations.

The only thing that counted in Germany's rearmament in the air was the will of the political leaders. This was represented by Hitler whose personal intervention in planning was felt more and more as the war went on. Even Göring, who had to represent the Luftwaffe, was primarily a politician. The person responsible to him for the execution of the air rearmament programme was the Quarter-master-General (Air). At the outbreak of war this post was occupied by Udet, as Chief of the Technical Department of the Air Ministry.

When the war with Poland began in September, 1939, Germany enjoyed complete air superiority. In the absence of any adequate Polish fighter defence it was easy for the German fighters to achieve their since well-known successes, and Göring was quite right when, speaking in 1943 of the Polish campaign, he said: "There was no *tour de force* there."

The same thing happened in 1940 in the expedition against Denmark and Norway. Once again there was no fighter defence to speak of. Yet the success of the campaign depended upon the Luftwaffe, as Narvik could be held only with the help of considerable air transport. Germany was then the only warring power with these transport fleets at its disposal.

In the West the blitzkrieg against Holland, Belgium and France was again to demonstrate the superiority of the German Luftwaffe. But in this case the bomber squadrons were beginning to become dependent on the escorting fighters if they were to avoid heavy losses.

41

In these campaigns the superiority not only of the Luftwaffe but its material as well had been made quite plain. It was the quality of that material and the high standard of the crews which ensured mastery of the air. They formed the protective roof under which the operations of the army could proceed undisturbed. The complete success of the opening moves had seemed to confirm the views and expectations of the German leaders. Yet the Battle of Britain which began on the 12th August, 1940, was soon to show that the German Luftwaffe was unable to win mastery of the air with the material available to it. Göring subsequently explained this failure himself:

"It soon appeared, however, that the British, in contrast to our Luftwaffe which was designed for the offensive, had even in peacetime prepared the Royal Air Force primarily for the defence of the British Isles. That also explains why so little had been seen of the Air Force previously. I remember the early night propaganda flights of a few individual aircraft in the Rhine area in 1939 and 1940, with a single bomb dropped from each machine. That was the sum total of English aggressiveness. Even in the Western campaign against France for all practical purposes the R.A.F. was absent from the battle against our army and Luftwaffe.

"But all the greater emphasis had been laid on the defence side and the force had been built up accordingly. Above all, they created an ideal system of radio detection. The detection of German aircraft became unfailing and the British fighters, directed and controlled from the ground, could be led easily to them."

The Battle for Britain was broken off in April, 1941 because the preparations for the campaign in the East were in full swing.

On the 7th February, 1940, before the blitz campaign of that year, there was a conference to consider the acceleration of the armament programme. Göring presided and Colonel-General Keitel, Colonel-General Milch and Reichsminister Funk took part. The report of this conference reads:

"Field-Marshal Göring laid down as an essential principle that without regard to what had been the practice hitherto the fullest use must be made of all stocks of raw materials so that the greatest possible quantity of finished products can be turned out as soon as possible.

"*Only those projects will be considered absolutely essential which will be completed in* 1940 *or promise to be producing by* 1941 *at the latest.*

"All other long-range programmes are to be examined again. Of special importance is it to concentrate on the vital items in the armament programme, as we in the Luftwaffe have done by renouncing the development of individual types."

This embargo on further work on individual types was based on a narrow interpretation of an economy demand of the Quarter-master-General (Air) of the 7th February, 1940:

"The shortage of aluminium and other non-ferrous metals forces me to the following conclusion: I consider it absolutely necessary that in the immediate future deliveries of the main types in use on the active fronts must be maintained at the highest possible level. It is my view that a reduction of the deliveries of aircraft for employment behind the front, such as training aircraft, which can be replaced by converted fighters or 'destroyers', can be accepted for the immediate future. Such a change in the programme would result in shifting the centre of gravity of the delivery programme in favour of aircraft types in actual use at the front."

So in view of these facts—shortage of raw materials and the ban on new designs—the main task of the German aircraft industry was to produce the front-line aircraft of existing types and improve those on the priority list. A strict ban was also imposed on the independent experimental work by individual firms. Acting on orders from above, the Reich Air Ministry strangled the creation of new prototypes which it called secondary, "as such types will not be wanted after the war." To realize the enormous significance of this decision, one must remember that in 1940 the war had hardly begun and was far from being won, and that the development of an aircraft from drawing board to serial production took three or four years, that of an aeroplane engine four to five.

After the beginning of the campaign in the East, when it was already obvious that the strength of the Luftwaffe—with the material and resources then available to it—was not equal to the further tasks imposed upon it, this ban on experiment was confirmed in another "Führer Order" of the 11th September, 1941:

"The armament industry is overburdened with requisitions and the fulfilment of the programme I have laid down is possible only if the demands of the services are harmonized and adapted to its capacity. The programme I have laid down therefore calls for the following measures:

1. the requisitions of each of the services must be strictly limited to their essential requirements;
2. the O.K.W.[1] must decide priorities when the requisitions do not comply with the conditions set out above;
3. the most careful examination of the requirements of the services having regard to the capacity of the industry.

"To make certain that these measures are carried out I therefore order that requisitions by the services should not be sent to the productions departments except through the head of the O.K.W.

[1] High Command of the Armed Forces.

The latter will discuss with the Minister of Armaments and Munitions the possibility of the armament industry's meeting them and he will, under my authority, decide the nature and scale of the orders to be passed. (Signed) Adolf Hitler."

The fiat of the O.K.W. dated the 10th October, 1941, laid down that the following should be the priority programme:

1. production of aircraft;
2. the Flak programme (Luftwaffe and Army);
3. the Flak munitions programme (Luftwaffe and Army).

With regard to experimental developments the decision was as follows: "After joint investigation with the Minister of Armaments and Munitions of the financial and technical feasibility of the requirements or proposals, the head of the O.K.W. will decide whether and to what extent these should be met or carried out. Only when that decision has been given may work be begun on them."

By an order of Göring of the 8th January, 1942, these O.K.W. powers were cancelled as far as the Luftwaffe was concerned and the State Secretary and the Inspector-General of the Luftwaffe received the following instructions: "The development programme of the Luftwaffe is to be reconsidered from the point of view of its practicability in view of the present position of raw materials and industry."

To this order of the Führer must be attributed the fact that the campaign against the Soviet Union which opened on the 22nd June, 1941, was not decided in a few months—as he expected—but in increasing measure swallowed up resources which the military leaders wanted devoted to army needs. The Luftwaffe formations engaged against England were only supposed to be employed temporarily in the East. This appears from a subsequent summary by Göring:

"And then came the Russian campaign. At that time I still cherished the hope that we need not break off the campaign against England. At that time it was a question of this or that Bomber Group staying only four days, just to give the impression that it had been transferred to the East. But what went East did not return. It stayed there."

After the victorious conclusion of the "Eastern Blitz", i.e. by the winter of 1940-41 at the latest, it was intended that the Luftwaffe should make available larger contingents, industrial capacity and labour manpower with a view to a general increase of production. Hitler said he could remedy a shortage of industrial manpower by releasing up to 500,000 men from the army. His calculations proved faulty. The factors of time and space were and remained forgotten.

Meanwhile the German aircraft industry continued to produce

the types of the first war years, viz. the Heinkel He 111 and Junkers Ju 88 bombers, the Junkers Ju 87 dive bomber, and the Messerschmitt Me 109 and Me 110 fighters. Greater speed was intended with the fitting of more powerful engines and a higher payload with the increase of wing surface area. But in practice it turned out that speeds, and more especially range, actually diminished, owing to additional demands for more and heavier armament and armoured protection, as well as larger bomb loads.

From these five years the German High Command did not draw the conclusions which were necessary for total war, particularly as regards the choice and production of material for the Luftwaffe. As the war went on, in view of the certainty of American support for the Allies and the possibility of German preventive measures against Russia, our military leaders were bound to have to face the question whether future development, and more especially production, for the Luftwaffe should not be given the highest priority in the general scheme of things.

In the summer of 1941, under the influence of the increasing British air raids into Germany, Udet had uttered a warning: "If our fighter arm is not substantially increased and we have not turned to the offensive by 1942 the war will be lost." Here our experience in the Battle of Britain should have opened our eyes. The essential preliminary was to realize that so far the employment of modern technical resources had shown that the Luftwaffe would be the decisive weapon, whether Germany remained on the offensive or was forced on to the defensive. Nor must Russia be underrated in that connection.

These technical problems for the future conduct of the war, to be solved as a preliminary to all other tactical, strategic and even political measures, were ignored. The technical aspect was pushed on one side as a nuisance; its requirements were rejected as being pessimistic, and realization of the growing armaments of the Allies was laughed at as being exaggerated.

In August, 1941, the Heinkel He 111 bombers had to be withdrawn even from night operations in the west as enemy night fighters made their losses too high. The range of our bombers was no longer adequate, and their armament proved too weak. The flying time of the Me 109 was also too short for them to escort the bombers on their raids and give them sufficient protection. Moreover the outbreak of war with the U.S.A. promised even higher demands on the Luftwaffe, to which it could hardly prove equal, whether industrially or from the manpower angle.

Even in this critical phase of the war the Luftwaffe General Staff's requirements from the armaments industry revealed a lack of clarity in its judgement of the situation. Throughout the war that authority never got down to really full-scale comprehensive strategic planning. The decision as to numbers required, which must always be made six to eight months before serial production could begin, never had in mind the way in which the war could be expected to develop. In September, 1941, for example, the General Staff ordered only 360 fighters a month for the year 1942. Milch, on his own initiative, increased that figure first to 720 and then to 1,000 a month. Before November, 1941, no new designs were available. The designs of the pre-war period and the first years of the war still held the field. Until the end of 1942 the aircraft constructional programme still contemplated the continued production of bombers that were designed for offensive use, but had meanwhile become outmoded.

No doubt this was a result of our too easy victories in the air war against Russia which blinded the German High Command to the true military value of the Luftwaffe. The great bulk of our aircraft was retained in the East and not withdrawn, as intended. As a result of the low quality of the Russian Air Force we enjoyed a long period of superiority on that front. Even the slow Ju 87 Stuka greatly distinguished itself. Even so, it was no longer possible to make good all our losses.

After the entry of the U.S.A. into the war the German leaders seemed to have been smitten with blindness where the prospective development of the Anglo-American air power was concerned. By the beginning of 1941 President Roosevelt had established a central production department (O.P.M.—Office of Production Management) and with the whole of America's industrial capacity made available corresponding results were to be expected in every theatre of war.

In fact, the entry of the U.S.A., with its effects in the material sphere, was to decide the war. The active intervention of the 8th Army Air Force bomber groups, the strength of which rose from twelve aircraft at the start to a total of more than 700 bombers, with corresponding fighter escort in 1943,[1] completed the collapse of German industry. Göring's prophecy that America could perhaps make Fords and Chevrolets but could not make aeroplanes had proved false, but he did not admit his error until the end of 1944, when in the same breath he said quite seriously that the U.S.A.'s reserve of men was now exhausted, and so the American air raids would soon cease.[2]

---

[1] Figures given to the author at Glucksburg on the 15th June, 1945 by General Fred Anderson, of the 8th Army Air Force.

[2] More than 193,000 men were trained as pilots in the U.S.A. during the war.

Up to the time when Milch took over in the autumn of 1941 the Quartermaster-General (Air) had had engineers in top posts to advise him and see to the execution of his orders. But now Göring issued an order that the top technical positions in the Reich's Air Ministry were to be held by General Staff Officers, and officers with front-line experience. The latter had of course proved themselves in action, but had had too little technical training to master the complicated workings of the aircraft industry. Even after this reorganization the following outstanding defects remained:

1. Lack of technical foresight.

It is easy to understand that soldiers in general can only develop on lines they understand, i.e. compatible with their previous front-line experience and not based on acquired technical knowledge.

2. Disregard of the most elementary rules of a rational production.

(a) The fundamental principle of all rational production is to select the most promising from a number of competing designs and then manufacture from that design alone. In the German Luftwaffe one looks in vain for a single case in which that elementary rule was observed. In the effort to meet all the demands of the forces, even the smallest special request was fulfilled and a very large number of small lines, with idiotic subdivision, was produced.

(b) Complete confusion took the place of rational order. Everyone made everything. It might have been expected that each of the existing aircraft construction concerns would be given a special task within the framework of the whole scheme, e.g. Messerschmitt, fighters; Junkers, large bombers; Heinkel, medium bombers; Arado, transport aircraft. Instead, the great ambition of each firm was to be represented in every branch of aircraft construction. Messerschmitt built not only fighters like the Me 109, Me 262 turbojet fighter and Me 163 rocket fighter, Me 110, Me 210 and Me 410 (fighter-bombers and night fighters) but also totally different prototypes such as the Me 321 Gigant glider, the Me 323 powered version of the Gigant, the Me 264 as a long-range reconnaissance and bomber aircraft. It was the same with Focke Wulf: the Fw 190 fighter, the Ta 154 night fighter, the Fw 200 long-range reconnaissance bomber, the Ta 400 as a rival design to the Me 264, the Fw 189 army co-operation aircraft, the Fw 44 trainer, and so forth. It is superfluous to cite the examples of Junkers, where the variety of different types was even more terrifying. It was the same story with Heinkel, Arado and Dornier.

All this was equally true of every other section of production. Take the aero motor industry, for instance. Instead of choosing one engine out of all the designs submitted and then putting it into large-scale production we simultaneously manufactured from three equally good prototypes, the Bayerische Motor Works BMW 801, the Daimler Benz DB 603 and the Junkers Jumo 213.

It is almost impossible to imagine what increased production, with what economies in materials, machine tools and labour we could have had if the basic principle of serial production had been observed, and what an advantage such a radical simplification would have been from the angle of supply lines and delays at the front. The usual objection that it was not wise to stand on one leg because one was thereby exposed to enemy action did not hold water. Quite the opposite. Production of only *one* type did not imply manufacture in *one* place but production of the same model in different places.

3. Irresolution and lack of logical thinking.

There was hardly a decision that was not reversed several times and then finally restored. As an example, no one could ever decide whether to stop manufacture of the Ju 290 (design and serial production approved), the He 177 and other types. A decision was considered at conference after conference but was always postponed. Meanwhile a monster staff of constructors and the most highly-skilled operatives were left working at top speed, and the preparations for mass production of some particular aircraft were allowed to go on, even after several decisions not to proceed to manufacture, so that millions of working hours and a vast quantity of expensive parts and specialized apparatus were wasted before a conclusion was reached that the type should not be produced at all.

Göring, under the pressure of events and in the teeth of experience both at home and abroad, now believed that he could soon close the production gap he had created by his own order of the 7th February, 1940. On the 12th September, 1942, he said: "The time required for the production of an approved or new aeroplane, from drawing board to serial manufacture, is such that we can only hope we shall never get it because we shall get peace first. I've never understood why that should be so. I am told it cannot be helped and so I suppose I must accept it. But it is monstrous that a new engine or fuselage takes years before it can be used."

The result of this official opinion was even more undue hastse in designing and serial production. Experimental aircraft which had not been tried out were put into production and there were several cases of aircraft types being put into production and proving so unsuitable at the front that the model had to be scrapped.

Udet, the first Quartermaster-General, and his successor, Milch, had tried hard to bring about a simplification of types and in part succeeded. In addition to training and naval aircraft and gliders, the following prototypes were built:

1939:   17 prototypes with 3 variations
1940:   14   „   „   6   „
1941:   16   „   „   7   „
1942:   22   „   „   6   „
1943:   23   „   „ 10   „
1944:   27   „   „ 11   „
1945:   15   „   „   6   „

This failure and the obligation to create and develop new aircraft and the continuous call for "modifications and refinements" in prototypes under construction put an unhealthy strain on designers and our whole industrial capacity. The explanation for the shortcomings was partly due to erroneous directions from the General Staff which in turn were to be attributed to an exaggerated idea of the capabilities of the technicians. In many cases loads and performances were called for which subsequently proved utopian. The results of these false expectations were that too many new developments were put in hand on the assumption that sufficient industrial capacity would soon be available. This led to dissipation of effort, ever-recurring bottlenecks and ultimately chronic postponement of delivery dates.

This was a permanent condition in the Messerschmitt company. Here the highly-valued chief designer, Professor Willy Messerschmitt, had to direct and control a huge and complicated industrial concern and, being hopelessly overworked, he could not give full rein to his outstanding creative ability as a designer.

A further unfortunate corollary was that both the top engineers and others less eminent were far more interested in new designs and improvements than in something which was often more important—testing what was coming from the production line.

In the year 1942, forty models were being developed in addition to the twenty-two aircraft prototypes with six variations. That year the Messerschmitt company alone had eleven new and modified developments of all sizes on the stocks.

On the 12th September, 1942, Göring commented:

"I've had to put up with a lot of hard knocks in the last year. Aircraft which were to do wonders, and the subject of great promises, have miserably failed to come up to scratch. The output of other aircraft which they were to replace has thereby been seriously retarded and in four cases I have had to go back to the earlier aircraft and step up their production because the new one did not turn out to be what was promised. This applies both to fuselages and to engines."

The "offensive" spirit which animated the Luftwaffe leaders up to 1942 expressed itself in a call for fast bombers and excellent diving performance. If possible, the bombers must be as fast as the enemy fighter, if not faster. These requirements hindered the production of a really effective bomber, not to mention the development of a suitable escort fighter such as the Western Powers had brought on the scene, while constantly extending their range. The progressive diminution of the opportunities for employment, and the corresponding ineffectiveness of our medium bombers both by day and night, had been established in the first year of the war. The fact that they were less obvious against Russia almost to the end of the war was due to the poor performance of the bad Russian fighters or the superior German fighter escort—a position which immediately changed when English or American fighters appeared in large numbers on the Eastern Front.

In the autumn of 1942 even the Dornier Do 217 and Junkers Ju 88 bombers had to be withdrawn from night raids against England. At that time our Western enemy's night fighter was the Bristol Beaufighter with a speed of 315 m.p.h. at 14,000 feet. In the absence of suitable aircraft the bombing war against England had either to be called off or carried on at great loss with "improvised" aircraft.

In 1942 the standpoint of the Luftwaffe leaders was as follows: "For aerodynamic reasons higher performance to-day can only be attained by increasing the performance of the power unit. A new aircraft must have a more powerful engine. But the same possibility lies in the further improvement of existing types; so a new aircraft design will not mean any progress. Increased engine performance alone will make possible heavier armament as well as bigger loads and increased range."

The countless programmes set forth by those responsible for air armaments suffered from the shortage of raw materials and industrial capacity, not to mention technicians and labour, allocated to the Luftwaffe. The heads of the armed forces regarded the air arm as an auxiliary weapon in the land and sea war. The demand that the Luftwaffe programme should be given priority within the framework of the provision for all the services was rejected over and over again. The army equipment programme—tanks and anti-tank weapons —which Hitler, as Commander-in-Chief of the Army, promoted, and the U-boat programme were treated as more urgent up to 1944.

It should be added that it was the Minister for Armaments and Production, Albert Speer, or his representative, Saur, who dealt with Hitler personally on the requirements of the army and navy. They put their views and wishes before him almost daily. Often enough the Luftwaffe was not represented at all, as Göring did not like technical discussions at the Führer's H.Q. So the air arm slowly became more and more discredited. Even when the American daylight raids became increasingly violent from the summer of 1943 onwards, there was little change. The men at the top could not get it into their heads that the speedy creation of a powerful defensive air force had become a matter of life and death.

Field-Marshal Erhard Milch of his own initiative now tried to increase the output of fighters after certain proposals to that end by Jeschonnek and Göring had been turned down. But owing to limited resources his efforts failed as he naturally could not and would not give up his bomber production. All these burning questions were thrashed out in many conferences under his chairmanship, attended by all the high air armaments authorities, which were part of the so-called State Secretary Meetings, held every Tuesday and Friday in the Great Conference Hall of the Air Ministry. The participants were heard and could speak their minds freely. Shorthand typists took down every word and the transcript formed the basis for planning the work to be done. In course of time he proved himself a master of debate. At these conferences he pulled out every stop of his undoubted rhetorical talent, his gift for taking everything in at a glance and his powers of organization.

But he was not always quite careful about the truth when giving reasons for his decisions and he did not often stick to one opinion long. At a programme conference on the 15th October, 1941, he said: "I consider the programme too far-reaching, especially when we consider the military position which means that the great campaign against Russia will be ended in the near future."

A year later, on the 5th December, 1942, he said something else: "Everyone knew there would be war in the East. It was long before June. I was asked whether we should prepare for the winter or not. I at once gave the order: Get everything ready for the winter. The war in the East will last several years. I know the East and its immense distances. I've often been there. I've knocked about the world a lot and know well enough what the risks are. . . ."

There was no real increase in production until 1943–44 when, for reasons to be given later, the full strength of our air force could not be employed.

The building-up of the Luftwaffe before the war was deliberately based on the assumption that it was to be used offensively. At a meeting of Gauleiters and Reichs Commissioners on the 8th November, 1943, Göring had this to say on the subject:

"At the start of the war Germany was the only country in the world to have a strategic air force with machines that were absolutely modern from the technical point of view. The other states had split their air force into an army air force and a navy air force, and above all considered the air force as a necessary and important adjunct to the two other services. For that reason they lacked that instrument which alone can deliver concentrated and shattering blows—the strategic air force. In Germany we worked from the start on the lines that the main body of the Luftwaffe should fly deep into the enemy's country and operate strategically, while a detached portion should primarily appear on the battlefield as dive bombers or, of course, fighters."

It was widely believed that the protection of the home country could be left to the anti-aircraft guns in conjunction with a few fighter wings. The general view was that the flak artillery with its guns and sighting apparatus would be in a position to prevent invading aircraft from employing aimed bombing tactics at any rate, and inflict such losses on our opponents that the raids would end of themselves. We also hoped that by continuous air attacks on their bases any air action against the German homeland would be nipped in the bud. The events of the first war years seemed to justify that view. The German aircraft industries, which had been moved from the endangered areas to the Baltic coast in particular, returned to their old quarters after the victory in the West.

The opening of Anglo-American air activities against the homeland towards the end of 1941 created a very different situation. The necessity for a powerful defensive force for the protection of our armament industries was fully realized by Udet and Milch, but they were not successful in convincing the Wehrmacht leaders. Up to the end of 1943 Hitler and Göring remained unshakeable champions of offensive war. No doubt in the case of Hitler, with his insistence on the maintenance of the "retribution" attack on England even long after the Reich had been forced on to the defensive, ideological and political considerations were uppermost in his mind. The ugly facts of the general military situation were secondary to the question of prestige. This was more than bluff. It was the result of that belief in his own genius and the infallibility of his own decisions which had been born of the victorious blitz campaigns and was nourished on propaganda and the flattery of the men around him.

Against this official view of the highest authorities Milch renewed his efforts to overcome the numerical weakness of our fighter force. His first plan (spring 1942) had provided for raising the output for that year to 1,000 fighters a month. But Jeschonnek, Luftwaffe Chief of Staff raised objections: "I cannot find hangars for more than 400 to 500 fighters." Galland's plan for a "Mammoth Fighter Force", which took account of all the European and North African areas then occupied by German forces and provided for a strength and monthly production of 4,000 to 5,000 fighters, was received with laughter and treated as lunacy. After many exciting conferences, Jeschonnek was persuaded to approve a programme of 700 to 750 fighters per month, i.e. less than Milch had himself proposed for the end of 1942.

Milch intended a further acceleration of the fighter programme in the course of 1943. He hoped to have an output of up to 2,000 a month at the beginning of 1944 and increase it to 3,000 in the summer. In all probability, his secret thoughts were fixed on 5,000. But with the existing production capacity at his disposal he could not get his programme accepted, even though it also provided for an output of 750 Ju 188 and 200 He 177 bombers for the summer of 1944. His fighter figures had still to suffer from the sullen resistance of Hitler and Göring.

Meanwhile, from February, 1944 onwards, the Western Allies' systematic air onslaught on our aircraft industry was beginning to take effect. The official requirements were never met.

The "failure" on the technical side of air armament had now become obvious even to Hitler, and the reaction was typical of a totalitarian régime: Göring, whose prestige with Hitler was unlimited until 1942, fell into disgrace. The suicide of Udet and Jeschonnek had been the start. The heavy raids on Hamburg in August, 1943 "finished" Göring, even in the eyes of the nation. Yet he remained in his many posts. What the continuous representations of recognized experts and a Udet and a Milch had failed to accomplish was now achieved by the heavy Allied raids on German cities and industrial centres. The highest authorities now realized—too late, of course—that without a powerful defensive air force for the protection of the Reich our war industries and our transport would be annihilated and the war would be lost.

On the 1st March, 1944 the "Fighter Staff" was established, responsible to Saur, Supervisor of Armament, who was given dictatorial powers as the representative of Reichsminister Speer. The aircraft industry was taken out of Milch's sphere and largely ceased to be a concern of the services in order that Saur could convert it into an integral part of the totalitarian state.

But even Saur could not get the much-desired simplification of the production tangle. The men around him came—like Saur himself—from labour circles which knew nothing of the Luftwaffe. There is a difference between mass-producing tanks and turning out aircraft in great quantities. The decisions of the new men were often based on the information and advice given to them by those who had been failures in the Air Ministry and industry. In other cases they made up their minds without concerning themselves with the requirements of air warfare. There was no end to the mistakes, as Saur came to no well considered conclusions and never found, or wished to find, a really competent air adviser.

Saur's ignorance of the nature of air power, which he shared with the political leaders, despite all the drastic demonstrations given by the enemy over German territory, is best shown by his remarks at a Fighter Staff meeting, on the 8th April, 1944: "We cannot win the war with our airmen but they are a prime necessity for the creation of conditions under which tank production can go on. It will be the tanks with which we shall win the war in the East."

It was only after the Anglo-American invasion in the West and the heavy enemy air raids into the homeland which concentrated on the aircraft factories that we really got down to fighter production. Simplification of types was again a factor, though this programme still contemplated 32 different types, to be reduced to sixteen for 1945–46. On the 8th July, 1944 Hitler approved the proposals and ordered that production should cease of all aircraft which we could do without at that stage of the war.

A short time previously the turn which the war had taken forced an important organizational reform upon us; air armaments were at last assigned to the Minister for Armaments and War Production.

It is hard to say whether this reorganization in 1944 under Saur, who continued to represent Speer, would have proved a real success, as to all intents and purposes Germany had lost the war by the end of 1944. But it is certain that Saur and his organization, which did not include a single strong man from the Luftwaffe, could hardly have revolutionized our aircraft production. It was not just a question of increasing output; leadership, with ability to plan ahead and take long views, was just as important. The general war situation, particularly the intensified bombing by the Western Powers and its effect on the whole industrial structure of Germany, would have called for measures to which military and technical knowledge were an essential preliminary; the new men did not possess it.

The following were the indispensable conditions for success:

1. *Ruthless simplification of types in airframe and engine production.*

In view of the heavy enemy bombing attacks an increase in output could be made possible only by concentrating on quite a few *outstanding* airframe and engine types. The manufacture and storage of these types should have been confined to well-distributed and— as far as possible—safe sites. The selection of those types should have been most carefully considered and then immediately acted on. The entire German aircraft industry should have been employed on the production of those types with the tactical aim of destroying the Allied escort and bomber swarms. The increase in production brought about by such radical concentration on a few types would have been very great if the supply, storage and repair apparatus had also been simplified. Manpower could have been released and immediately employed in removing the factories to the new sites and making them really safe against attack. Work on types abandoned should have stopped *at once* and the transfer to the selected types proceeded simultaneously with the cessation of production of the former. The hiatus in production could have been kept within reasonable time.

2. *Protection against air attack.*

The German aircraft industry did not understand the necessity for immediate protection from air attack and was criminally late in realizing it. Ruthless action was called for. The factories which could not go underground, for lack of time and labour, should have been dismantled and, if necessary, transferred to barracks and woods. This would have held up production for a time, but concentration on a few types would have brought about a general increase in output.

3. *Strict adherence to decisions once made.*

This requirement should have been fundamental. The slightest modification in mass production spells ruin. It was possible to get first-class work from unskilled labour when thousands of aircraft could be produced without alteration. But if modifications were required immediately after mass production started, the result was a run on the trained experts and eventually a slowing down of production.

4. New developments in aircraft and jet research should have continued independently of the general programme but even here been confined to a few types. But large-scale production of a new type should not have been *in addition* to existing mass production of older types; the production of one of these older types should have been stopped, so that large numbers of the new type could be produced.

The monthly output of fighters mounted steadily and in September, 1944 reached its maximum of 3,129, including the first rocket and turbo-jet aircraft the Me 163 and Me 262. These aircraft actually went into production but their fighting value did not correspond to their number.

At the end of 1944 it was no longer possible to talk of a systematic direction and control of industry in Germany. The last attempt to maintain adequate armament production, in the face of conditions dictated by the enemy, was the appointment of "Armament Commissioners" by Hitler in the spring of 1945. Its head, SS-

Obergruppenführer Kammler, behaved as a self-willed dictator who was not responsible to anyone. It was an act of desperation which in the last hour unnecessarily cost lives and wasted material.

Looking back on ten years of military rearmament, it can be said without fear of contradiction that in all technical branches almost superhuman feats of pioneering had been accomplished. It was these achievements of German research workers, inventors, engineers and industrialists, and the devoted intelligence and self-sacrifice of the German working man which made it possible for us to wage a mighty war against the world for six long years. The technical advantages which German aircraft undoubtedly possessed at the beginning of hostilities were amazing, considering the circumstances in which they were obtained. They can only be fully appreciated when we remember how the development of our aircraft industry was shackled from 1919 to 1933 by the embargoes and limitations imposed in the Treaty of Versailles, while the industries of all air-minded countries enjoyed state aid of every kind which enabled them to gain a big start in design and experience.

For the fact that Germany in the short interval before the war was able not only to catch up with them but actually leave them far behind we owe thanks mainly to the knowledge and ability of a number of men active in aeronautics and industry. Nor must we forget that the concentration of all means to one end, a process partly based on that ideological unity in the Third Reich for which Adolf Hitler strove, provided a powerful if short-range impetus. If the lead we had won from the democracies was not maintained during the war, it was primarily due to the very rigidity of the totalitarian system, though our limited resources contributed. Our leaders stuck to the outworn strategic notions of mass warfare and it never occurred to them to answer the Allies' "quantity" with a shattering blow based on "quality". It was not the German nation and its greatest brains which failed us in the long trial of war but the false and rigid *"Führerprinzip"*.

Post-war efforts to invent "cases of sabotage" and thereby account for our loss of the war was absurd. It is not by such means—beloved of our eastern neighbours—that we shall arrive at a sober judgement of the collapse of our aircraft industry. It is significant from every point of view that from 1943 onwards the actual mass manufacture of our aircraft was almost entirely left to foreign workers, and yet output did not fall and there was no sabotage, despite the enemy air raids. Throughout the war there was no sabotage worth mentioning in the German aircraft industry.

CHAPTER V

GERMAN AIRMEN AND THEIR WEAPONS

On the 6th April, 1936, I was admitted as a cadet to the Air Warfare School at Gatow airfield, Berlin. The syllabus provided for a careful training in all branches of army and flying service. We were one of the first cadet classes of the young German Luftwaffe. Out of ten thousand, only a few hundred had had the luck to be selected.

The moment we entered the establishment we had the impression that no trouble or expense had been spared. Our quarters, the sports ground, the gymnasium and swimming bath, the flying-ground, halls, lecture-rooms and aircraft were all the last word of their kind. The whole atmosphere of the place breathed the spirit which animated Hermann Göring, Commander-in-Chief of the Luftwaffe, in building up the new arm. We swore by the Führer and worshipped Göring as the greatest air hero of the First World War. For years we had spent our spare time and holidays learning gliding on the slopes of the Rhön, or at Rossitten, in the Borken-bergen, or on the Ith. Now the leaders of the new Germany had made it possible for our dreams to come true.

The first jolt to our young enthusiasm occurred during the pre-liminary military training. Our drill instructors were no airmen. We were handled on strict military lines. We learned to see Mother Earth from the frog's angle and had it impressed on us that thinking for ourselves could have lamentable consequences.

In fact this Flying School was quite unfitted to make independent and self-reliant individuals out of us. Flying officers in a position to grasp the complicated mechanism of a modern air force were not turned out in such schools. Was it surprising that before long there were "rebels", among them several who were to become the most successful airmen of the Second World War. Lent, Philipp and Claus were among them. We were the last to get through the commission-ing examination and I well remember the parting words of the Commandant: "I dismiss you with considerable doubts about your own careers and even greater doubts about the Luftwaffe itself." Two years later most of our crowd had either given their lives or won the highest distinctions.

Our ideas of flying had become a bad dream. We young flyers suddenly lost touch with the world around us. With every fresh experience of the heights above we began to apply fresh standards to men and things. We spoke another language, though we were anything but the soulless robots or bloodthirsty Huns of subsequent enemy propaganda. It was under the compulsion of human actions for which we were in no way responsible that we became agents of the murderous urge for mutual slaughter which seized the earth.

We flew and in the skies forgot the horrors of the war and the death and havoc we distributed by the push of a button. We had no voice in the earth-shaking events; our function was to fly and carry out orders. Yet the day came when we could speak out too.

For me that day came in August, 1941. It was then that as a young lieutenant I made my first attack on the hide-bound element within the bomber squadrons. At a meeting with the Reichsmarschall and his Chief of Staff, General Jeschonnek, I held forth about the necessary reorganization of the bomber force. Speaking for the fighting front, I insisted that every commander of a bomber group[1] or wing[2] must be capable of leading from the air and not from a desk. This sounded positively revolutionary and as I spoke Göring seemed to shed his years. He immediately agreed and at the end of my unintentionally long speech he promoted me to captain—a reaction I had not envisaged. Before the day was over, I went to Führer Headquarters with Mölders, the new Inspector of Fighters.

In Fighter Command there had been a progressive process of rejuvenation since the campaign in the West. The best fighter pilots were distinguished by the number of their kills. The top-flight men at that time were Molders, Galland, Wick and Osau. The rapid rise in their rate of kills was accompanied by an equally rapid rise in rank. Within a few months they became wing commanders. Anyone who could not maintain a high rate of kills would not remain a squadron leader for long. Thus the front line units were led exclusively by "aces" whose ambition was, and must be, to keep ahead of everyone else in their personal kills and keep their squadron ahead of any other. They were not yet tactical leaders of large formations though the position changed later on with the growth of the enemy fighter and bomber arm.

The bomber crews looked at their job from a different angle. Outwardly they seemed less excitable, tougher and more self-possessed—attributable to their wide-ranging sorties far behind the enemy frontiers, often over water and in bad weather at night, and perhaps most of all to the slower speed. But as the war went on greater flexibility was required of the bomber pilot in the control of his aircraft, particularly in attack and defence.

[1] Group: about 120 aircraft, known in German as a *Geschwader*.
[2] Wing: 30/40 aircraft, known in German as a *Gruppe*.

On their sorties in bad weather or in fighter-bombers the fighter pilots learned what was required of us so that towards the end of the war a more or less standard type of front-line airman was evolved. Yet a good fighter type was seldom a good bomber type and vice-versa. The latter must always be on guard not to be shot down. However desperate or intricate his evasive measures, he must never forget his primary object of placing his bombs in the centre of the target.

In contrast to the fighter force, the bomber squadrons did not offer a composite picture in the first years of the war. Until the year 1941, with few exceptions most of the bomber squadrons learned little from their sorties which had not been taught them in training. The He 111s and Do 217s usually dropped their bombs in level flight, whether flying in groups by night or singly by day. This helped the old staff officers, often in the company of a general, to show their prowess in their "Knights Cross Course". But in fact a young lieutenant or captain often took over the actual leadership of the squadron after the start. Of course there were exceptions.

It is safe to say that the bomber squadrons had achieved their original successes in the "blitz" campaigns without operational or even strategic direction. Tactically speaking, the commanders of the aircraft and squadrons were the lieutenants and captains, whose mentality, way of life, military outlook, youth and personal influence with the crews were in such sharp contrast to the parade-ground, heel-clicking atmosphere surrounding the starched dug-outs set over them that as the war went on an unbridgeable gulf separated the two camps into which the flying service was split. Fortunately the "staff majors" were on the side of youth. These were the war veterans who had retained their affection for flying even in their peacetime occupations and had kept themselves young in heart. What should we have done without them?

In the course of 1940 further squadrons were equipped with the Ju 88. In this connection the "old gang" proved particularly tough and unyielding. Their great standby was a most respectable but excessively formal and fossilized general who was about as suitable for the post of Adjutant-General of the young Luftwaffe as Methuselah for Cleopatra. So once my mind was made up I applied direct to the Chief of Staff who had remained young in heart and was always on the side of youth despite his forty years. It was through him that my report reached Göring.

It was the direct result of this intervention, which coincided with a similar representation from Captain Pelz, that at the end of 1941 a command training school for bomber pilots was started. Pelz was put in charge. He was one of the most gifted among us and Göring gave him unprecedented authority. Henceforward no one could command a flying formation who had not successfully passed through this school. Even the "old gang" had to pass the test. The wheat was separated from the tares and this command course became the seed-bed of a stock of young leaders inspired with hostility to everything which was not "front-line", the High Command of the Luftwaffe, the staffs of the Air Fleets, the Ministry of Air Transport and the "Party hangers-on". Yet they were intensely loyal to the Führer and their Commander-in-Chief. Göring, who in good times had done everything in his power for his airmen and on the human side often displayed a positively infectious affection and concern for his men, would have found this band of young front-line officers his unfailing support even in bad times.

While the rejuvenation of the fighter and bomber forces was going on there was a simultaneous reorganization at the top. After Mölders met his death in a crash, Galland was appointed Inspector of Fighters. At the end of 1942 Pelz and I were moved from the Mediterranean to Berlin as Inspectors of Bombers. Our sphere of responsibility covered training, the tactical and technical application of the lessons learned at the front and appointments to positions of command. We were also given a certain right of veto *vis-à-vis* the General Staff by the Commander-in-Chief of the Luftwaffe.

But the fundamental opposition between the front and the home command could not be allayed. The difference between their functions in war is admittedly so great that only men with the highest personal and professional qualities could hope to bridge them. Moreover, as the war went on and the area of operations progressively extended and the enemy's superiority became ever greater, the demands on the front-line formations were harder and harder to meet. "Star" performances by individuals became of even less importance in bomber operations. In a night sortie in bad weather who could say which crew had aimed best?

Thus the ideas and practices prevailing in the First War had been transformed. In those days the fighter had been the principal agent in air warfare, but it had now to surrender its rôle to the bomber, which must now provide the backbone of air strategy. Influenced by the theories of Douhet and Rougeron, the Luftwaffe leaders adhered to that view both during the creation of the new arm and throughout the war. The designers and technicians produced their bombers with that view in mind and were convinced that it was practically invulnerable as well as superior to all foreign aircraft. Yet, even though front-line experience demonstrated the exact opposite in the early days of the war, the Luftwaffe leaders and the overwhelming majority of the official experts were not to be shaken in their opinion that they could win the war with this tried and trusted weapon.

During the Second World War only three types of bombers, the He 111, the Do 217 and the Ju 88, in addition to the Ju 87 dive bomber, were employed by us on a large scale. On the morning of the 1st September, 1939, we started with the He 111 against Polish targets and in April, 1945, we used the Ju 88 to launch the last guided bombs against the Russian bridges across the Oder. The only twin-engined bomber with which successful dive-bombing could be effected was the Ju 88. It is fundamentally true that the German Luftwaffe carried on its air offensive with aircraft of types evolved and established before the war and which had but a very limited range.

In the last war the bomber always proved inferior to the fighter. This statement will prove equally true for the future, even if new inventions may give the bomber the upper hand for a short time, perhaps as the result of higher altitudes and speeds.

German aircraft production, particularly as regards bombers, during the war was determined mainly by the short-term requirements of the individual land theatres. Those requirements, which were often conflicting, and the absence of a strategic bomber fleet, always stood in the way of systematic, long-range planning, development and production.

Some considerable time before the war the General Staff had called for a medium bomber with a range of about 900 miles and a bomb load of 2,200 lb. in order to have northern France, including Paris, within range. This requirement was met only by the He 111H; the Do 17 M, also developed from a civil aircraft, could at first carry no more than one 550-lb. bomb.

With these aircraft a great deal of information about bombing in level flight was acquired in training and during the Spanish Civil War. It had been established that that type of attack was only feasible for big targets. Nothing less than a stick of bombs offered much prospect of a hit. At least four 550-lb. bombs were required for industrial targets. As it was still not possible to employ a complete formation of Wing or Group strength in bad weather, high-level attack was only possible in good weather. With the twin-engined aircraft of those days and at that stage of engine development greater range plus bigger bomb load seemed an impossibility. So experience appeared to show that diving attacks at an angle of 20 to 30 degrees offered the best prospect.

The general staff now extended their requirements by asking for a new medium bomber with a range of 2,000 miles, a speed of 250 m.p.h. and a crew of four, suitable for diving at up to 30 degrees.

As early as December, 1937 the Technical Department was in a position to inform the Air Staff that these requirements seemed to be met by the Ju 88. This aircraft, one of which had been adapted by Junkers as a racing type, set up a record in 1938. In that flight from Dassau to the Lugspitze and back the Ju 88 had attained an average of 300 m.p.h. It was generally known as the "Wonder Bomber".

The Ju 88 was equipped with two Jumo 211 engines—the same as for the He 111. So the improved performance was not obtained by any change in the engine but by reducing the wing surfaces. Objections to the increase in weight (empty) were silenced after practical flying tests. On the 3rd September, 1938, Göring gave Koppenberg, head of the Junkers works, *carte blanche*: "Go ahead and give me a great bomber fleet of Ju 88s in the shortest possible time!"

Although the aircraft industry made a tremendous effort it did not prove possible to reach the prescribed output. Mass production started at the beginning of 1939. The first squadron was equipped with Ju 88s shortly before the outbreak of war but it was the late summer of 1944 before they could be supplied in quantity. But the aircraft had not been thoroughly tried out and a large number of teething troubles showed themselves in action at the front. They appeared at the front without effective diving and night flying aids. When these became available at the end of 1941, dive-bombing was nearly out of date. During 1941–42 the engine performance had to be stepped up and the wing surfaces extended, to compensate for a bigger bomb load and additional equipment. Yet even this improved Ju 88 could no longer appear in the West in the daytime. Further modifications had no better result. The aircraft did not prove much faster as was intended and, as it was impossible to install supplementary petrol tanks to meet the increased consumption of the more powerful engines, the range was being continuously shortened.

In this case, as with other types, German air armament adopted the fatal policy of trying to maintain the original technical superiority of German aircraft by constant minor modifications. It was impossible.

At the beginning of 1943 the Inspector of Bombers reported as follows on the usefulness of the Ju 88: "The present Ju 88 is to be regarded as no longer fit for service by day in the Anglo-American theatre by reason of the extremely heavy losses: speed, armament and endurance are inadequate to cope with the enemy fighters and their escort system. Fot the time being it is serviceable for night fighting if we discard two bomb racks and any speed-reducing armament and equip them with airborne interception aids. On the Eastern Front it can be used in good weather by day, and at night when the losses are not unreasonably high."

Thereupon Jeschonnek, the Chief of Staff, ordered the production of a suitable twin-engined bomber to take the place of the Ju 88 and the Do 217 (which was open to the same criticism as the Ju 88) which were being employed against England. The growing enemy defence and the failure of our escort fighters which tried to protect the German bombers on their raids called for the creation of a fast bomber.

Even in 1941 the Luftwaffe leaders had been reduced to the expedient of equipping fighters such as the Fw 190 with bombs for use against England. They first carried bombs of 550 lb. and then up to as much as 2,200 lb. But this half-measure solution known as "Jabo Raids" caused our fighter forces losses out of all proportion to the results obtained. Towards the end of 1942 fighter-bomber raids on England were renewed on Hitler's personal order as part of his "retribution" campaign against the British Isles.

The first fighter-bomber raid was a surprise and a success, but the British were better prepared for its successors and there were heavier losses which, having regard to the other claims on our fighter force, could not be borne, so that this type of raid had soon to be called off. But the Americans copied and developed it with their long-range Mustangs and Thunderbolts.

All the attempts of the Junkers and Messerschmitt concerns to evolve from their proved models a serviceable twin-engined fast bomber such as would meet the Air Staff's requirements must be regarded as unsatisfactory. Neither the Me 410 nor the Ju 188 and Ju 388 fulfilled the optimistic hopes of their designers. The main obstacle was still a satisfactory solution of the engine problem.

In the search for a suitable aircraft for the renewed "retribution" onslaught on England which Hitler had ordered, the new "Leader of the Attack on England" Colonel Pelz, on a visit to the Dornier works stumbled on an existing design, the Do 335, which the Technical Department had some years previously rejected as unsuitable mainly on the ground that escape by parachute was difficult and uncertain.

Realizing that the designs for single- and twin-engined aircraft, which had to a certain extent become standard in all countries, could not produce much higher performances, Dornier had long been interested in a tandem arrangement of the power units. The losses incurred through the engine cutting out convinced him that he was on the right track. After years of preliminary theoretical and practical work an aircraft with one power unit in the nose and another power unit in the tail was designed. When the call for a serviceable fast bomber came in 1943 he at once offered his Do 335 to the Technical Department. The fact that the power units were in the nose and in the tail meant that the wings were free from encumbrances while still of moderate size. With less wind resistance a greater range and speed could be obtained without more powerful engines. The aircraft could be flown by any average pilot. If the forward or rear engine cut out there was no danger, whether in flight, taking off or landing. It could fly on one engine.

Pelz was able to interest Göring in this machine. In November, 1943 immediate production of the Do 335 was ordered. By the beginning of January the first three prototypes were ready. Unfortunately, the heavy Allied raids on aircraft works, one of which was Dornier's at Friedrichshafen, held up the first serial production so that up to the end of the war only eleven of this type went into service. This aircraft was the last German attempt with piston engines to use "quality" to overtake the enemy's technical lead and ever more menacing superiority in resources.

Still searching for the fast bomber for which the high-ups were insistently pressing, all eyes were now turned on the long familiar subject of rocket propulsion.

From the middle of 1943 Hitler and Göring had begun to take an interest in jet fighter aircraft, and particularly the Messerschmitt Me 262. Hitler soon raised the question whether it could not be used as an extremely fast bomber. On the horizon loomed the preparations for a large-scale invasion by the Western Allies. Many of Hitler's remarks showed that the possibility of its succeeding was worrying him very much and from midsummer of that year he took his countermeasures. One of them was the introduction of the jet bomber.

Hitler did not pay too much attention to the danger literally overhanging the capital in the shape of the Allied strategic bomber offensive. His main concern was that the fronts should hold. So this new weapon, which would have been the best answer to the enemy bombers and their escort, necessarily fell victim to the Führer's obsession with the idea of offensive war. His closest advisers, Göring among them, did not dare to oppose him, though they expressed other opinions in their conferences. The responsible leaders of the aircraft industries tried to gloss over their failures by optimistic and highly coloured descriptions and reports. At first Hitler swallowed them. But after ever-recurring difficulties and delays he had had enough of argument and towards the end of the war left the decision of the jet aircraft question to the S.S.

Hitler's concern, under the menace of invasion, to have suitable means of defence available had not grown any less by 1944. On the 5th January he impressed on Milch that he was anxious first about the new submarine and then the jet aircraft. "If I get them in time I

Werner Baumbach

The author talking with his crew before setting off on a bombing raid

In the cockpit of his plane

can beat off the invasion and the rest of it. Other things are important but not so vital."

Meanwhile the Allied air raids on German cities and the armaments industries were getting heavier. In view of the superiority of the Allied fighters, which accompanied the raids on Berlin, and the mounting weakness of our defending fighters, which had to put up with inferior material and poorly trained personnel, there was a rising chorus of voices among the fighter leaders: "The Me 262 is a fighter and its place is in the defence of the Reich!" The fighter Staff agreed with them, and so there was a historic conference, Hitler himself presiding, on the subject of the Me 262 at the Obersalzburg at the end of April, 1944.

At that conference Milch and Sauer in Göring's presence demanded the large-scale production of the Me 262 as a fighter. Against his better judgement Milch had said that this aircraft was not designed for bomb carrying, though the fact was that in 1942 it had been planned as a fast bomber. Its adaptation for employment as a fighter also had been called for subsequently by Fighter Inspection. Milch tried to compromise, but this time he had miscalculated and reckoned without Hitler's excellent memory. Hitler was extremely angry and said: "What I ordered has not been carried out." He gave a categorical order that the Me 262 should be produced as a "blitz bomber" at top pressure.

Göring, who also originally championed the idea of using the Me 262 as a fighter, now failed to voice it. In any event his influence with Hitler had sunk so low that it would have made no difference. From now onwards no one was more ardent in supporting Hitler's idea that the aircraft should be used as a bomber.

Göring's helplessness in face of a Führer order was clearly betrayed in what he said subsequently: "To exclude the possibility of error, we must not go on calling the aircraft a fighter but a fast bomber and also hand over the whole business to the General of the Fighters. The Führer's will is that some of the prototypes shall be given further trials and developed as fighters." General Bodenschatz interjected: "But the Führer strongly insisted that the experiments as a fighter should continue." Whereupon Göring replied: "Only the experiments! The prototypes must suffice for the production of the fighter while mass production is as bomber only."

Galland and Messerschmitt tried to convince Göring that it was absolutely necessary to use the Me 262 as a fighter, especially to cope with the Mosquito raids, for which it was particularly well suited owing to its high rate of climb. Göring was not open to argument. These were his final words:

"So that there shall be no misunderstanding I repeat that the

Führer does not wish the aircraft to be written off altogether as a fighter. But at this critical moment he wants it for other purposes. I hope to God that some day the Me 262 will become an outstanding fighter, and so does the Führer. But now the vital question is whether I have an aircraft which can really cope with the enemy when he comes at us with his whole air power, as he is now doing in Italy and will also be doing over the Channel. Let there be no mistake; the time for discussion and debate of the fundamental question has gone by."

The same day Göring issued a whole series of orders and instructions designed to remove doubts about the Me 262. He emphasized once more that the Führer had no objection in principle to a Me 262 fighter. But the fast bomber must be available first, even if it carried only one 550-lb. bomb.

The mass production of the Me 262 brought fresh difficulties, as it was actually designed by Messerschmitt only as a fighter. The armour and armament had to be incorporated, creating problems of weight and weight distribution for which allowance had to be made when bombs were dropped. Anyhow, even the Me 262 as a bomber arrived too late for the Allied invasion of 1944. They were occasionally employed in the late summer, but without particular success as they were not provided with suitable bomb-sights.

In October, 1944, after discussion with various front-line air commanders, Göring decided to call a general council of war. It was the sole occasion throughout the war in which the toes of the totalitarian concept of leadership were trodden on. Some thirty front-line commanders personally selected by Göring, met at the Air Force School in Gatow, Berlin. Their purpose was ruthlessly to tackle all problems of the direction of air warfare. The conference became known as "Areopagus".

After Göring opened with an assurance that we could voice any criticism except against himself he let Pelz, who was *persona gratissima*, preside and retired to Karinhall to devote himself as he said to all these burning problems. A remarkable feature was the absence of the Chief of Staff of the Luftwaffe and other high officers responsible for the direction of the air war. As some compensation the Reichsmarschall's court camarilla was represented by his First Aide-de-Camp and his technical adviser. A full shorthand note was taken.

Many reasonable proposals were discussed but there was also much bandying of irrelevant arguments betraying a complete lack of sober judgement of the situation. In practice this "Areopagus" ended in chaos. When it was over and I went to Karinhall with the fighter pilot Colonel von Malzahn to give Göring the transcript, he merely asked me whether I myself would have demanded the removal of his personal entourage as a preliminary to any reorganization of the Luftwaffe. I replied that I would.

That very afternoon he decorated his chief Aide-de-Camp, Colonel von Brauchitsch, with the "Air Leader's Gold Medal with diamonds" and assured him of his complete confidence. He declined to commit himself with regard to the practical proposals. That was the end of Göring's prestige in the service.

After this last effort I deliberately by-passed Göring and had direct discussions with Albert Speer, the Reichsminister for Armament and War Production. Last moment though it was, we wanted to employ what was left of our bomber force in a heavy strategic attack on the Russian power plants. The preparations for this operation were covered by the code word "Eisenhammer".

As the military collapse of Germany in a few months appeared to be inevitable, we thought that such a blow might favourably affect the outcome of the war because the advancing Russian steamroller would be cut off from its sources of replenishment. Thereby we might even stay the progress of the Soviets, though the effect of such an onslaught would not be felt at the front for three or four months.

Operation "Eisenhammer" was to be carried out by Bomber Group 200, which was under my command and had become the last reservoir into which all the aircraft weapons and special accessories of the bomber force had been poured. Among them were the "Father and Son" aircraft which were to carry out the direct attack on the power plants. As far back as the 17th January, 1942, Pelz and I had drawn Göring's attention to the idea of these aircraft, which were a technical achievement of the Junkers works. Unfortunately, without result. The idea was of a crewless Ju 88 above which a fighter such as the Me 109 or Fw 190 was attached. This coupled contraption, which was also known as Beethoven, was flown by a pilot seated in the fighter. Instead of a crew in the cockpit there was four tons of an explosive which was more devastating than anything that had previously been used in air bombing.

The double aircraft approached its target in a glide and when the pilot had the latter in his sights he set the guiding apparatus and uncoupled his fighter when he was about 3,000 feet away. While the Ju 88 with engines running pursued its guided course to the target, the pilot was the faster off the mark, thanks to his additional gliding speed. He was running comparatively little risk. As the fuel had been previously pumped out of the Ju 88 the range of the fighter was considerably extended. The accuracy of this weapon which though new was assembled from existing aircraft, was about eighty per cent against fixed targets. Its effect was shattering.

Technical misunderstanding of air warfare and petty jealousy of all suggestions put forward by "outsiders" delayed any further developments for years. Operation "Eisenhammer" never came off because East Prussia, the proposed base, fell into Russian hands too soon. In March, 1945, more "Father and Son" aircraft, with increased range, were forthcoming for operation from the base at Berlin. Then it became obvious that the Anglo-Americans would leave Berlin and East Germany to the Russians and so it would have been madness to go through with the plan. It would probably have meant that hundreds of thousands of German prisoners of war would simply be set to work to rebuild everything we had destroyed. So we sent the last of these aircraft against the bridges which the Russians had thrown across the Oder. That was the last exploit of the German Luftwaffe.

CHAPTER VI

THE "BLITZ"

"WHAT about Hitler's Blitzkrieg in the West? Tell us about 'the Blitz'. "

I was always being asked that question when I was spending a short leave in Japan after returning from Narvik. The Japanese General Yamashito, Commander-in-Chief of the Army Air Force and subsequently conqueror of Singapore, was a particularly attentive audience. Hitler's "blitz" had staggered the world. Poland, Denmark, Norway, Holland, Belgium, France—all knocked out in a few months! How had he done it?

The Luftwaffe had been the roof under which the army's land operations could develop unhampered. It had also been a tactical auxiliary in the land fighting. The German soldier, his morale, his fighting spirit, his arms and equipment, had in land and in the air proved far superior to the Western enemies.

When the German bomber squadrons penetrated far into Poland in the grey dawn of the 1st September, and German fighter and destroyer aircraft gained complete command of the air in a matter of days, neither we nor the world in general had any idea how and when this Second World War would end. Hitler meant to win the "blitz" campaigns, whether his generals and General Staff officers wanted to or not. The German soldier and Germany's youth were solidly behind him, believed in him and were his willing slaves.

After the destruction of the Polish armies and the occupation of Warsaw, the German armed forces enjoyed a considerable breathing space lasting up to February, 1940. In that period German bombers of the He 111 and Ju 88 types began to attack warships in British harbours and off the English coast. Based in north-west Germany and the Frisian Islands in February, 1940, these attacks were extended to the lair of the Home Fleet in Scapa Flow. The best crews were selected from the different bomber crews to form the Lion Bomber Group 26 and the Eagle Bomber Group 30. Even if the results of these first raids were to some extent exaggerated by German propaganda, the fact remains that the British fleet could no longer feel itself safe from attack. The dive-bombing attacks on Scapa and in the Firth of Forth have become milestones in modern air warfare.

In conformity with an official declaration of the German Government on the outbreak of war, Hitler issued strict orders that on these

Luftwaffe sorties not one bomb should be dropped on land targets
of any kind. At any cost the German flyer was not to be the first to
start indiscriminate bombing from the air. This Führer order, which
was drummed into the crews every time we went out, depended on
observance by both sides, as the Governments of Great Britain and
France had declared their adherence to the relative provision of the
Washington Agreement of 1922. On the 15th February, 1940, the
British Prime Minister, Neville Chamberlain, emphasized the
importance of this clause, which all the belligerents had agreed to
respect.

Minor attacks, such as British bombs on Esbjerg and Western
Germany, were ignored and not followed up by Luftwaffe reprisals.
It was only in the weeks of crisis caused by the German offensive in
the West, that the R.A.F. began dropping bombs indiscriminately
on German cities. The starting point was a raid on Freiburg im
Breisgau and Heidelberg. The military value of these raids, which
were carried out on an utterly inadequate scale, was precisely nil.
From an historical point of view they can only be regarded as acts
of desperation. But the effect on Hitler and public opinion in
Germany was certainly enormous. The German reply was not long
in coming.

The battle with England was now joined on the basis of un-
restricted air war and on the German side was begun with a shower of
leaflets: " 'Churchill has attacked Heidelberg. That's our answer!' "

Hitler's right to "retribution" is today admitted by such eminent
British military writers as General Fuller and Captain Liddell Hart.[1]

Meanwhile, we had had first successes in sinking a heavy cruiser
with a direct hit and damaging or sinking some merchant ships and
other warships. The naval war had become land and sea war.

Extract from my diary at that time:

"Off we go over the North Sea in the direction of Andalsnes.
The moment we leave the coast we are wrapped in a dense snow-
storm. We are soon flying higher. In another half-hour, our target,
reported as a group of enemy warships, must be reached.

"We get everything ready for attack. In the last few moments
the tension is tremendous. I'm always asking my navigator: 'Aren't
we there yet? We must be there!' Then we've suddenly arrived,
recognized the target and yelled out as one man: 'a battleship!'
Voice of the gunner: 'Below us a big battleship and quite big
transport round it!'

"From now on we do not let the big 'tub' out of our sight: at
first I think it is a battleship myself; it looks so big in the narrow

[1] The author says nothing about the German attack on Rotterdam which could
certainly not be justified as an "act of retribution"! (The German Publishers).

fjord. But when we get close I clearly recognize the striking and characteristic form of a cruiser. As clouds cover the target at the same moment I cannot deliver my attack but have to turn and run in again.

"There is fierce A.A. fire from the cruiser and the transport. On this next run in I discover that my neighbours have discovered a hole in the clouds and are diving down. I can observe the fall of the bombs and see that they are near but too short. The A.A. fire seems to be growing weaker. I know I must drop my bombs on the second run as the weather is fast getting worse. Otherwise the cruiser will get away.

"We are running in. The cruiser tries to escape by zigzagging. But this time we surprise it by appearing through a tiny gap in the clouds. We rush down and I keep my ship in the sights all the time. We mustn't miss!

"Bombs away. The aircraft gives a heave of relief when it has got rid of them. At the same moment the gunner shouts: 'Hit midships to starboard!'

"Our machine-guns hammer at the A.A. guns on the transport. There is pandemonium. I step on it and the jagged mountain tops come nearer. I wrench the aircraft just past them. The gunner feverishly takes photographs. Explosions, smoke and flames have hidden our victim, which a few minutes later begins to sink. After a last look round we disappear in the clouds.

"We fly back. We radio our success. As we sweep into our bay on landing I can see the Group Commander's car coming towards us. On the tarmac I make my report.

"It is 7.40 p.m. and we go off to supper. As we enter the dining-hall we are told that in the 8 o'clock news there will be a special announcement telling the whole world of our successful attack. For the first time in history a cruiser has been sunk from the air."

To secure our northern flank in our intended attack on the West the occupation of Denmark and Norway seemed to the German leaders to be a necessity.

That bold gamble, the occupation of Norway, came off. Once again it would have been impossible without the Luftwaffe. Glider units and air transport columns, in conjunction with the dive-bomber crews, were the arteries through which the marine and moun-tain troops, fighting on land, were tirelessly kept going and through which the success of the operation was guaranteed.

On the British side the lack of unified leadership and the absence of any fighter defence may have contributed materially to the speedy end of British aid to Norway. On the German side the limits to what our own Luftwaffe could do were plainly to be seen. But there was no time for reflection, let alone application of the lessons of the Norwegian campaign to the equipment and training of airmen. Even before the English had left Norway, part of the Luftwaffe formations employed in Norway had already been transferred to their starting bases in north-west Germany.

The 10th May, 1940, was the opening day of the German victory in the West. Fort Eben-Emael, the Albert Canal, Rotterdam and many other names became milestones in the headlong onrush and glorious memorials to the German parachute and glider troops, who captured apparently impregnable bastions in bold *coups* and, supported by the airmen, thus laid the Belgian and Dutch back areas open to the advance of the army divisions following up.

I may quote again from my war diary:

"On the 9th May I learned from the wireless that I had been awarded the Knight's Cross. Next morning we are out again at 3.30 a.m. The night of the 10th May has brought the decision. The vital thing is to capture and hold certain key positions in Belgium and Holland until the army formations crossing the frontier can come up. The *coup* planned cannot succeed unless there is very accurate timing of the operations.

"Our job is to attack flak (anti-aircraft battery) positions in the Rotterdam–The Hague area. A pitch-dark night hides all preparations. After a long time small calibre bombs are fixed. Our fine technical chaps work feverishly while we, the flying crews, at once allow ourselves a few hours' more sleep. Take-off at 4.30 a.m. is our orders. . . .

"We fly into thick mist and the contours of the Dutch landscape below us become vague. The engines hum their monotonous song and for a moment we are all wrapped in our own thoughts. It is the rhythm of war, modern air war, which drones in our ears."

"We are abruptly brought back to reality. Our battle area has been reached. Salvoes of white puffs accumulate ahead of us. The mist prevents us from recognizing our target as yet, but it is also our protection, as at the same moment my gunner calls out: 'above to the left, enemy fighter.'

"I know them, these Dutch aircraft. I swerve to the right and dive my bomber down into the mist. We have to protect ourselves from unpleasant surprises. . . .

"We are in touch with the ground again and must be close to our target. Ahead of me the leading aircraft are already diving. I soon recognize the exact position of the little target on the edge of the airfield; it is a flak battery firing from a farm. It has a commanding position on a little hill. We must hit it if the landing of the parachute troops coming up behind is not to be endangered. I can already see the first bomb burst on the southern edge of the airfield.

"I decide to come in from the West in a wide curve and make my attack from a low height. I do a steep glide and then dive. The target rushes up towards me and the farm in my sights gets bigger every second. I don't notice that my left thumb has pressed the little button on the stick. 'Bombs away,' reports my gunner, followed by his dry 'direct hit'.

"As we turn away we see that the farmhouse has collapsed like a pack of cards. The flak is silent. We fly back. Hardly were the words 'now keep a sharp lookout for fighters' out of my mouth before we are attacked by a Fokker G.I. Thies, my wireless man, has spotted it in time. We fire away to keep the thing off and I search desperately for a protective cloud. Nothing! Our only salvation is 'dive like the devil!' I hear a voice: 'damnation, my gun's jammed!' But Willem Braun behind me has got his machine-gun through the perspex above. I take a quick look round. The enemy destroyer is very close indeed. I can easily see the pilot. Then everything is drowned in the clatter of our two machine-guns. Thies is firing again. We dive to safety. The destroyer does not follow us. An Me 109 has got it.

"At the same time we see our paratroopers and glider men go in to the attack. They have arrived to the dot.

"On the return flight we could see the field grey columns of our army surging along the Dutch roads."

Throughout the Western campaign the tasks of the Luftwaffe up to the final defeat of France were confined within the framework of the overall strategic and tactical design. Here was begun what afterwards became stock procedure in Russia—the army automatically calling for air support before any operation was attempted.

Colonel-General von Brauchitsch and von Rundstedt had objected to Hitler's plan for the Western campaign, particularly because part of the German concentration plans had fallen into the hands of the enemy. The men around General Halder seriously considered putting Hitler under guard to avert what looked like inevitable disaster. But the miracle came off. The German infantry and armoured divisions swept forward to the Channel in an irresistible flood. The strongest fortifications and mightiest obstacles yielded to their youthful frenzy. The spirit of Maginot collapsed under the mighty blows of German bombs and tanks' shells.

The great lesson to be drawn was that fortifications constructed solely to meet frontal attack cannot hold if the supply lines of the defenders, and the morale and activity of the civil population behind them, are worn down by continuous attack from the air.

The main problem of the German High Command throughout the Western campaign was how to keep the ever advancing army groups supplied. Out in front, young tank commanders, company officers and wing leaders seized the initiative everywhere to keep the fleeing enemy on the run. Divisional and even higher staffs had few tactical or strategic worries in those days; it was all they could do to keep up with their units and find suitable quarters. Air formation had to keep a sharp eye on the movements of the ground forces lest they should endanger our own spearheads.

Hitler made headlong for Paris in the middle of the armies swarming over Belgium and France. In improvised "Führer Head-quarters" he interfered personally in the conduct of operations and decided the daily lines of advance and the targets. To the public the General Staff had taken a back seat. Their first scepticism vanished in the "Miracle in the West". Propaganda trumpeted the Führer's "military genius" far and wide. Most of the German people swallowed it. He believed it himself.

From now onwards his military advisers were nothing but "extras" in uniform, officials with nothing to do but carry out his wishes. He could promote, demote, replace, dismiss and hang them, or even—as in Field-Marshal Rommel's case—send hired assassins in generals' uniform to them. After the Western campaign he thought he could dispense with their advice. He needed them no longer except perhaps as a glittering suite.

No German emperor ever had as many marshals as Hitler had, yet none of them meant as much to him as the youngest lieutenant in the old Prussian army. The political dictator had become the infallible Commander-in-Chief. He entered Paris in triumph. "Europe for the Europeans," he declared in an interview with the chief correspondent of the Hearst press, Karl H. von Wiegand. "Will England negotiate with me now? I want peace with England. You know that I've always wanted it."

The German victory in the West was to prove fateful for both Hitler and, through Hitler, Germany. It had been no lesson, but was full of attractive snares of all kinds. The only lesson that was of vital importance to the Luftwaffe was never drawn.

The last military blitz operation was the Balkan campaign in March, 1941.

Our Italian ally had then been at war with Greece for several months and its divisions had been held up on the Greek–Albanian mountain frontier without forcing a decision. The German leaders thought that England would intervene with troop landings on a considerable scale. There was not only the danger that the Balkan Pact arrangement would be dissolved to Germany's disadvantage but that Italy itself would be directly threatened. Once in possession of the Dalmatian coast and the Straits of Otranto, it would have been easy for the enemy to roll up the Apennine Peninsula from the East. A secondary factor was fear of a Soviet invasion of the Balkans, despite Stalin's assurance to the German Military Attaché in Moscow a few weeks earlier that Russia stood by the Ribbentrop–Molotov Agreement of August, 1939. But who trusted whom?

The Balkan campaign, the result of Yugoslavia's change of course, was a preventive measure to ward off the threat to our southern flank and weaken England's position in the eastern Mediterranean.

Once again the appearance of the German Luftwaffe was the opening move. On the morning of the 6th April, 1941, shortly after Goebbels's broadcast justifying the war, the citizens of Belgrade were roused from sleep by German bombs. The campaign against Yugoslavia was brief; the deep national differences between Serbs and Croats were to our advantage. The Greek troops in their strong mountain defences offered obstinate resistance to the combined German, Bulgarian and Italian attacks. But our dive-bomber attacks smote the fortified works of the Metaxas Line and created breaches for the Axis troops pressing south to the Aegean Sea.

The German leaders were amazed at the absence of any real support for Greece from the British. The R.A.F. was tied down defending the home country against German bombers and the Middle East build-up was only just beginning. American "Lend-Lease" for England was not yet effective.

That was the way the blitzkrieg came to the Balkans. It was clearly shown that in the age of modern war from the air the sovereignty of any state is dependent upon the ability to defend and feed itself. Small nations like Poland, Norway, Denmark and Greece —and even Belgium and Holland—were helpless against air attack by a well armed great power. Fighting in isolation, they were beaten one by one in the absence of substantial aid from their stronger allies.

CHAPTER VII

THE BATTLE FOR ENGLAND

AFTER our successful blitz campaign Europe lay at Hitler's feet. There was only one enemy left—England. The outlook for an invasion of the British Isles seemed unusually favourable, but the army and navy insisted that the operation known as "Sealion", which was planned, must be dependent on the Luftwaffe first securing air mastery over the Channel and England itself, to prevent the Home Fleet from interfering with the crossing.

My notes, made at the time when the air war against England was in full swing but no decision had yet been made as to the form which future operations against her should take, may show how the position looked to us airmen at that time.

"The various sorties flown by our squadron against England are a fair reflection of the present phase of the war. Offensive action takes the form of air war and blockade. Air war is the affair of the Luftwaffe, blockade that of the Luftwaffe and navy jointly.

"Various bases, not too far from the British Isles, are available to the two services for their war of attrition against the citadel of the British Empire and its vital sea approaches. There are the Dutch, Belgian and French harbours on the Channel and the North Sea and Atlantic as well as their hinterland. In the Channel coast and the flanking position in Norway, the fourth shore of the North Sea, Germany possesses an extremely valuable defensive zone as well as an admirable base for offensive operations.

"A very large number of aerodromes, benefiting from greatly improved ground organization and protected by excellent A.A. defences, provide our bomber, fighter, reconnaissance and transport squadrons with concentration areas for the air war against Great Britain. The German operational air forces, thanks to their wide dispersal, are difficult targets for the British airmen. Admittedly their night bombers range far into Germany, but their efficacy is still low. Our proximity to the English coast which our victories have won for us places all parts of the country within reach. The rest is simply a matter of arithmetic—reduced flying distances, less wear and tear on personnel and material, less risk, less fuel to carry and so bigger bomb and mine loads. Efficiency has substantially increased as a result. Today enemy aircraft can carry to us only about one-third of the load which our bombers can carry to them from our enveloping coastal positions.

"England is also at a disadvantage on the armament front, a decisive front in total war. This island, situated on the edge of Europe, has to draw the bulk of its strategic raw materials, auxiliary supplies and foodstuffs from overseas. From a geographical point of view the island is placed most unfavourably. The majority of the military and important war industrial centres, as well as the ports, are in southern England or the Midlands. London has only just begun to realize the dangers of this concentration of its war industries, and is belatedly thinking of dispersing such targets so as to make them less vulnerable to air attack.

"England had no Nelson of the air in the two decades after the First World War, years in which the ardent theoreticians about the nature of the next war were working out how to convert the air force from the minor weapon it had been into an equal partner with the others. England too thought that the next war would begin where the first left off. Did she overlook the revolutionary changes which were taking place on the Continent in the military sphere as well as the political?

"The range of our aircraft has deprived England of her exclusively insular character. For the first time in history, the country is a battlefield, from Dover to the Shetlands and the Humber to the Mersey. The employment of bomber and fighter squadrons in the decisive battles in the West and the destructive onslaught on the enemy's supply and transport in the Western campaign are inseparably associated with the idea of the German Luftwaffe.

"But it is always the same. A man does something miraculous and most people will go on expecting him to perform miracles and are a bit disappointed when cold facts stand in the way of constant repetition.

"In the past few months has it not often been asked at home why our Luftwaffe had not yet laid England, the hereditary island foe, in ruins?

"We airmen can only reply: 'We know that England is the hardest nut to be cracked in this war. Our experience at the front has shown us that final victory against England can only be attained by the systematic co-operation of all arms of the service and ruthless application of the elementary principle of concentrating all one's strength and effort at the vital strategic point. This vital point is not necessarily the same as the enemy's strongest point.' Even if the air arm is the most important weapon in total war, it cannot *by itself* ensure the decision, final and total victory.

"Air war against England is in any event hampered by the

enemy's geographical position and all sorts of weather factors. We are separated by water. The first months of the air war against England have also shown that the British are tough and courageous opponents.

"The tasks now confronting the Luftwaffe are different in several important respects from those with which it was faced in the previous campaigns to strike England's continental weapon from her hand. The two main tasks of the bombers are to cut the island off from its overseas sources of supply and to obliterate the nuclei of the defences. Simultaneously our fighters must strive for command of the air over Great Britain. The strategic conception is simple, provided that the possibilities open to aircraft are clearly realized and our leaders think in strategic terms.

"In actual fact innumerable difficulties, great and small, have to be overcome and countless factors taken into account.

"Ideas have to be translated into action. Personnel and material, plans and their fulfilment have to be carefully fitted into an overall pattern. When that has been done, and not before, a reconnaissance squadron or bomber groups can be sent out with a definite task to perform. In this hard war against England the onlooker can have little idea of the effort demanded of our flying crews. He knows nothing of the sleepless, worrying nights when sorties have to be postponed again and again because the meteorological forecasters cannot promise favourable conditions.

"So our squadrons are carrying on with the blockade war. To us this is a pause between the campaigns while the land fronts are static. We have strewn mines in the river mouths and bombed ships and ports. When we return in the moonlight or the pale glimmering of dawn, reflecting on the sense and purpose of what we are doing and exchanging impressions of our experiences we all come to the same conclusion: we will never conquer England until the full might of the whole Wehrmacht is welded together for a knock-out blow and the full strength of the Luftwaffe is concentrated to that end."

Air superiority must thus be regarded as the first stage in the conquest of the British Isles. For that purpose how was the Luftwaffe placed, now that it was only 100 miles from London at the nearest point?

Even before the war we had, thanks to Colonels Rowehl and Beppo Schmitt, a useful target map of England, prepared as a result of reconnaissance flights by special units. But, apart from certain "optimistic" studies the General Staff had done little else in that direction. The task of destroying the Royal Air Force called for much more, above all large fighter forces with ranges covering

Southern England, including London. A wholesale offensive with bombers was doomed to failure unless these fighter forces could remain in the air over enemy territory long enough to bring the enemy up to fight and beat him there or destroy him on the ground. Without fighter cover, bombers, even when flying in dense formation, were easy meat for the British fighters. And the fighters of the Royal Air Force were good.

The vital question whether the German fighters could perform the double task of protecting our own bombers and shooting down the enemy fighters had to be answered in the negative. That was how our fighter aces on the Western Front saw it.

The flying time, and therefore the range, of the Me 109 was so short that only under the most favourable circumstances was an air battle in the vicinity of London possible for more than a few minutes. If our fighters and bombers missed their rendezvous at the exact moment or got separated on the approach, the whole sortie was wrecked and cost the bombers heavy losses. For one thing was plain —the Royal Air Force fully realized how critical the position was, and the English fighters had no hesitation in going bald-headed for our bomber squadrons.

In England a first-class radar system was already in operation. Its function was to pick up raiders and direct the English fighters and it was now having its first test. England owed this invention to the High Frequency experts Watson-Watt and Appleton. When our bombers flew in, the numerically inferior British fighter squadrons were always there to meet them and skilfully evaded our exclusively fighter sorties in order to save themselves for other occasions. But of course the resources of Fighter Command also were limited and if the German air offensive had been continued even the Hurricanes and Spitfires would have run out of breath. But with the help of their radio direction finding apparatus the English were able to conceal the actual strength of their squadrons and so compel the German command to break off the Battle of Britain.

The German fighters, which could not be directed from the ground, literally had their hands tied, tied to the Continent by an invisible but none the less potent thread—the question of fuel. Over and over again they had to return to base before the enemy could be caught and held. When action was actually joined the German fighters usually showed themselves to be superior. But the Royal Air Force had the advantage of fighting on interior lines and made wonderful use of it. As the British have always been masters of improvisation and experts in expedients, here too they stayed long enough in the air to make short work of the German bomber attacks (and inflict heavy losses) by day and harass them at night.

The German Me 110 "destroyers", designed as heavy fighters to escort the bombers, were a complete failure. They were comparatively so slow and their flying qualities so inferior that they needed fighter protection themselves. After extremely heavy losses, the Luftwaffe's daylight attacks on airfields in Southern England and the Midlands, the southern harbours and London had to be abandoned in October. Radar and the few fighter squadrons had gained the day. On the German side, the effectiveness of British radar was not fully realized and no serious measures were taken to interfere with the wireless organization or destroy it from the air.

The German reports of victories, designed for propaganda purposes to conceal the failure of the Luftwaffe against England, were a very sore subject with our formations employed in the operations. When Field-Marshal Milch, as Inspector-General of the Luftwaffe, visited certain bomber groups which were based in Holland, Group 30 made no secret of its indignation. The squadron leaders who had actually flown in these operations bluntly told the Field-Marshal that it was impossible to produce the results required of them with the aircraft, bomb-sights and armament at their disposal, that the English fighters were just as superior to German bombers as German fighters were to British bombers and that night bombing could have no decisive effect, as we had neither bomb-sights for night operations nor enough bombers. Milch seemed grateful for our frank statement and said he would immediately seek a remedy.

He did produce a remedy. One Wing of the Group which had borne the heat and burden of the attacks on British warships and merchant vessels was broken up "as punishment for mutiny and defeatism". The officers were transferred and reduced in rank. A new staff officer arrived who had promised Göring that he would restore the efficiency of the Group in a fortnight if he were given command. His request was complied with.

Meanwhile the army and navy had gone ahead with their preparations. For weeks the question of an invasion of the British Isles was in the balance. Then Hitler sounded the retreat. Göring, who had said that he "would do it with the Luftwaffe alone", was authorized to wear England out from the air and join the U-boats in strangling her sea approaches. Field-Marshals Sperrle and Kesselring were to use their air fleets to finish the task which had been thought too risky for a joint operation with the other arms of the service.

We had failed to secure control of the air over England and the idea of invading that country was given up for the time being. Hitler's thoughts turned to Russia. Instead of bombers German

(*Above*) A photograph taken from Baumbach's plane during a dive-bombing attack on a British cruiser during the Norwegian Campaign

(*Right*) Attacking a British battleship during the same campaign

(*Above*) Baumbach introducing Hitler to members of his Unit

(*Left*) Hitler congratulating the author on his exploits

fighters were now sent as fighter-bombers on daylight hit-and-run flights to England. After a few sorties, which proved very expensive, that method of attack was abandoned.

Night bombing of industrial centres, military installations, airfields and shipping and mining of the English East Coast were now to be adopted as the method of weakening resistance and playing havoc with supply. But the Luftwaffe effectiveness had been very highly tried. Weather conditions—fog and other navigational difficulties, meant that the bombing war in England in the winter of 1940–41 could only be effective against big ground targets. The idea of economic air warfare was substantially watered down. London, Liverpool, Portsmouth, Hull, Coventry and many other cities, as well as the East Coast ports as far north as Edinburgh, were singled out.

Hitler talked about "extirpating" the English towns and propaganda coined the word "coventrizing" for the maximum degree of destruction which was deemed to have been inflicted on Coventry.

In this night bombing during that winter London figured pre-eminently in the German communiqués and reports on the progress of the Luftwaffe's "wearing-down" strategy—which was not in fact achieving its purpose.

After a period in hospital following a forced landing in Holland on my return from a sortie to London I was back at the front in January, 1941.

The mining of the Thames estuary and the Wash, and attacks on shipping on moonlight nights, were continued. Night bombing of various targets was intensified, though the English fighters and A.A. defences were continuously improving while the efficiency of the Luftwaffe was rapidly decreasing. Yet though the counter-measures were not yet able to prevent the German raids, the squadron leaders, who a few months earlier had been accused of defeatism and few of whom had survived, had stuck it out.

At the beginning of 1941 the Luftwaffe High Command had to admit that England could not be reduced to impotence solely by night bombing, carried out as it was by inadequate forces. The mining of English waters had no great success as the enemy produced a simple and effective reply before our magnetic mines were available in large quantities. The Mediterranean theatre and the preparations for the Russian campaign drew off men and aircraft from the West. The Luftwaffe leaders had to abandon their hopes of another great concentration against England.

The German air war against her had failed. Neither the firs phase of the Battle of Britain, the fight for mastery of the air in the

6

summer of 1940, nor the second phase, the strangulation and destruction of the British economy, had been fought by the Luftwaffe
with sufficient resolution and in adequate strength.

Our target planning was both too casual and too periodic. For
weeks on end the big target of London was attacked, then the
Midlands reappeared in the time-table, together with feeble
isolated attacks on Liverpool, Glasgow and other northern ports.
It was not too difficult for the skilful British defence to anticipate
our current monthly programme and concentrate its available
fighters and A.A. accordingly. It frequently succeeded in deceiving
us with dummy targets and fires. "Don't shit, hit!" was a vulgar
expression frequently used by commanding officers in those days, but
it all boiled down to a few bombs here, a few mines there, and then a
quiet interval for the enemy in which to clear up the mess and
recover his breath. The battle over England cannot be compared
even remotely with what was to happen to Germany from 1942
onwards.

But it must not be forgotten that Germany had no established
pattern of air warfare to follow. Experimenting though she was, even
the expensive and abortive sorties taught her the secret of tactical
success and she became willy-nilly the pacemaker in the strategy of
the air. The Allies themselves, learning from our failures, proved
apt pupils of the Luftwaffe, apart from the fact that they simply
took over or developed innovations such as the employment of
fighter-bombers which we had found useful.

CHAPTER VIII

THE BATTLE OF THE ATLANTIC

WE were eventually to realize that the "coventrization" of the British Isles and the increasingly expensive attacks on the London area would not force the British to their knees. The German leaders were letting their own air force knock its head against the strongest bastions of the island empire, while the enemy could concentrate his available force and put up an ever more effective resistance.

Yet the real lesson of the German naval operations against England in the First World War had been that the British could be defeated only if the Motherland could be cut off from her overseas sources of supply. Her dependence on such sources was the Achilles heel of British world power. The small German navy was far less capable of that titanic task than the Kaiser's navy had been. It was only with the strongest possible support from the Luftwaffe that our navy could have any prospect of success in a struggle with Britain's superior sea power. But the conditions precedent to such co-ordinated sea-air warfare were not established by the German leaders in the period when the Wehrmacht was being created.

Göring, in agreement with Hitler, represented the view that the build-up of the Luftwaffe must not suffer any division of purpose in matters of organization, training and the development of aircraft types. The navy, on the other hand, believed that it should have its own air arm: "the section of the Luftwaffe selected for the sea war must be considered as part of the sea-going fleet and be organized and trained to co-operate with fighting ships. The crews must receive careful and systematic training."

It was to be otherwise. The Luftwaffe's practical preparations for air war at sea were limited to providing the navy with the necessary reconnaissance aircraft and a small number of specifically naval aircraft. The idea of a separate fleet air arm or special units for air-sea warfare was rejected because the Luftwaffe leaders were convinced that the decisive naval operations would take place outside aircraft range. Anticipation would be impossible, and everything would happen so quickly that effective air support by the Luftwaffe would not be feasible, even if it were within range.

Another reason was that at the time when the air force was being built up, aircraft engines were not considered reliable enough for use over water, even though land aircraft were regarded as superior to naval. It was not till 1938 that long-range flights with land air-craft over water were inaugurated. Though the results were satisfactory they had no direct effect on the fundamental ideas of air warfare. Neither before nor during the war did the German navy succeed in demonstrating the invalidity of Göring's opinion and possessing its own air service. To the navy air reconnaissance was the first consideration. At a later stage Hitler gave his own views on the questions involved:

"Taking the long view, I remain convinced that we cannot have two Luftwaffes. Strategically speaking, the Luftwaffe is the most effective weapon in conducting war as a whole. No one is in a position to say that he will fight with land aircraft on land and with naval aircraft at sea. I must start from the assumption that land aircraft cannot fly over water nor naval aircraft over land.

"Where should we be now if we had carried out that separation! We would not have been in a position to use the Luftwaffe strategi-cally. That we are is the great merit of Reichsmarschall Göring, who has made it his predominant aim and has always said: 'At first sight it seems splendid that each service should have its own air arm—a strategic Luftwaffe, a Luftwaffe for the army and another for the navy. But when you get down to realities, all possibility of concentration has been lost. Concentration may seem a simple matter. In theory no doubt it is, but we all know how difficult it is to get a naval battery, an army battery and a flak battery under one hat even if they are guarding the same small spot on the coast. How much more difficult with the Luftwaffe!'

"We should never have been able to hold our own in this war if we had not had an undivided Luftwaffe. There can be no room for doubt. If a critical situation arises anywhere, in the West for example, we can bring over everything we have from Norway—but not if what we have in Norway is naval aircraft. They could not be taken away.

"If we want to concentrate at one point we must mass everything we have at that point and make ourselves weak elsewhere. I should like to know how that would be possible if we had three different air forces, $33\frac{1}{3}$ per cent strategic, $33\frac{1}{3}$ per cent army and $33\frac{1}{3}$ per cent navy. In the long run it will be plain enough that the Luftwaffe, the most effective strategic weapon there is, only makes sense when it is so organized as to be controlled and directed by one hand.

"Now we are in a position to say that wherever a critical situation arises we can immediately concentrate to meet it. Today, when we are facing superior opponents, we can only hold our own by occasionally contriving to be quite weak everywhere else and scrap-ing up everything we have got."

Hitler's remarks were no doubt based on his appreciation of the basic problem during the whole period of Germany's armaments build-up. Shortage of materials, and still more of time, imposed limitations. Yet these should not have led to one-track strategic thinking, and the requirements in the way of arms and equipment which resulted from it. If the German leaders had rightly appreciated the political and military developments they would have taken good care that at least some part of the Luftwaffe was adequately prepared for air-sea warfare. An aircraft crew which could hit a warship travelling at speed would obviously be capable of hitting an aerodrome or a railway yard. Crews with experience confined to sitting targets and navigational training barely adequate for ground warfare, could not be used at sea, so that it was not enough to prepare several of the best fighter squadrons for sea operations just before the outbreak of war. Their numbers, training and arms were insufficient for their task. Their considerable initial successes must be ascribed mainly to the fact that their attacks on enemy shipping in the North Sea and Atlantic met with little or no opposition from enemy fighters or anti-aircraft. Within a year the position was to change. Even the employment of the four-engined Fw 200—a transport plane converted to a long-range bomber—had to be abandoned both against convoys and isolated ships owing to the increasing effectiveness of the anti-aircraft defence.

The first systematic, big-scale air operation against British sea-power was the aerial mining of harbours which was carried on from the winter of 1939–40 by squadrons equipped at first with the He 111 and then with the Ju 88.

By 1939 the development of the German aerial mine had progressed so favourably that its employment for military purposes could be envisaged. But orders for the mass production of this weapon were not given because not enough was known about its chances of success or the tactical method to be employed. None the less, the German leaders anticipated decisive results from the new weapon. The dropping of mines from aircraft would mean the systematic blocking of English harbours for a considerable time.

The Luftwaffe leaders considered co-operation with the navy unnecessary because aerial mining was limited to harbours and their approaches whereas naval mining extended to any open waterways used by ships. Aerial mining was also intended to supplement our bomb attacks on British ports, in the hope that harbours blocked by mines and sunken ships would become useless, temporarily at any rate, as channels of supply.

The conditions precedent to the anticipated effect of our aerial mining were the existence of appropriate aircraft and a vast number of suitable mines. Nothing less than sowing *en masse* could have any chance of success.

At the beginning of this mine-sowing warfare the Luftwaffe possessed two types—the 1,100-lb. air mine LMA 1/11, and the 2,200-lb. LMB 1/11. We had only 292 of each. Of these eighty-six were sown in England up to the end of 1939. With few exceptions the attacks were carried out only by daylight or on bright moonlight nights. The mines had delayed action fuses, and most of them were set so as to become "live" on the sixth day after being dropped. Mines without delayed fuses were also laid with a view to letting the enemy have a few cheap successes which would put him on the wrong scent. The mine could be laid at any depth desired. The parachute was released the moment it hit the water. The fuse was sensitive to the magnetic properties of the ship's hull.

In practice there were serious tactical and technical obstacles to the success anticipated. Mine-laying was dependent on the depth of water, weather, height from which the drop was made, the existence and effectiveness of anti-aircraft defence and other imponderables. The improvement of the British defences intensified the difficulties and very soon we had to give up operating by daylight and moonlight so that accuracy of laying suffered. The reports rendered by our squadrons did not always square with the truth and led the staff to have exaggerated ideas of the effectiveness of mine-laying by aircraft. They are revealed in a report of the 1st October, 1941, from Milch to Göring:

"The IXth Air Corps has informed the Reichsmarschall that up to the 31st July, 1941, 490 ships, totalling 918,000 tons, were sunk by mines laid from aircraft. At that time about 17,000 mines had been used, of which half must be excluded because they were dropped on land targets, so it follows that on an average every seventeenth mine sank a ship. The actual losses are greater because the above figure of sinking comprises only the losses of which Germany knows. But in addition to this destruction of enemy ships, his available carrying capacity has been seriously diminished by the continuous blockade of the sea approaches and the necessary provision of anti-mine auxiliaries. To take one example, a single sortie to the mouth of the Thames in which only 291 mines were laid closed that approach for sixteen days."

The optimistic views about mine-laying from the air led the IXth Air Corps to ask for a monthly supply of 2,000 mines, though that number was never used. The highest figure was 1,065 in September, 1940, when production was already 2,000. After 1941 aircraft for mine-laying were no longer available. This form of attack was continued for a short time with a handful of planes and then discontinued altogether.

The much exaggerated effect of aerial mining caused Milch to write to Göring again:

"Having regard to the success achieved with a comparatively modest outlay, the possibilities of aerial mine-laying should be exploited much more vigorously than hitherto. It is therefore proposed that immediately after the end of the campaign in the East the largest possible number of aircraft—certainly not less than 200—should be employed in mining English waters. As the A.A. defence at night is much weaker off the coast than on land, there should be an initial attack on the largest scale before it is strengthened."

The course taken by the war supplied the answer to this letter. The anticipated speedy end to the Eastern campaign did not materialize.

Further requests for an intensification of mine warfare with improved mines could not be met because the Luftwaffe no longer had suitable aircraft at its disposal. On this subject Admiral Dönitz addressed a memorandum to Hitler dated the 8th June, 1943:

"The use of the latest mine fuses offers us the hope and expectation of an urgently needed rise in the number of sinkings from mines. The effective employment cannot be attained by naval units alone; they must be supplemented by strong Luftwaffe forces in the areas which cannot be reached by ships. A determined mine offensive, carried out jointly by the Luftwaffe and the navy, should have outstanding results, particularly round the British Isles, in waters which at the moment are quite freely used by a great volume of hardly diminished traffic."

Some of our aerial mines were used against land targets, but the limited output of mines militated against systematic mine warfare from the air. As with so many other weapons of naval warfare, right up to the end of the war new and improved aerial mines were developed without ever reaching the production stage. German mine warfare never had any decisive effect on the operations of the Allies. It remained no more than a threat which was met in time by countermeasures which greatly diminished the success of the mines we laid.

For the time being the great initial success of the German submarines overshadowed the contribution of the Luftwaffe to the naval war against England. It was only when the British reaction drove our submarines further and further west that in the spring of 1942 our leaders remembered that in the Luftwaffe they had a useful weapon for attacking the well-protected Anglo-American convoys which were supplying not only England but the Soviet

Union through Murmansk and Archangel. At that time we had nothing in the West available for the purpose except a Ju 88 bomber squadron and an aerial torpedo squadron. Despite violent anti-aircraft fire, these weak forces achieved notable successes, though these were of little account having regard to our opponents' vast superiority in arms and equipment.

My diary records one of these sorties as far as the Faroe Islands:

"The sun is beating down on our airfield at Stavanger-Sola. My flight has been bathing in the fjord.

"We have been hard worked the last few weeks. Many tricky sorties, but some excellent results. Our crews have once again circumnavigated Scotland and shown that we can catch the enemy convoys anywhere. We have sunk enemy ships far out in the Atlantic, off the Hebrides and the Shetlands and the east coast of Scotland itself. Now we are much under strength. Every attack is a grim and desperate affair. The enemy is on his guard and victory calls for the utmost effort of aircraft and crew. It will be forthcoming.

"Today we have had a decent rest. The weather reports for tomorrow are favourable for the job I have in hand. Some big merchant vessels have been reported in the Faroe fjords. We will pay them a visit. It will be a long flight.

"I take off punctual to the moment. Our Ju 88 is literally fuelled to overflowing. The heavy bombs are stowed below.

"The Norwegian coast is soon out of sight and for hour after hour we fly over the wide Atlantic, our happy hunting ground today. We four members of the crew of our Ju 88 have been together throughout the war. We can rely on each other.

"Suddenly some dark objects appear on the horizon ahead of us. At first sight they look like aircraft, but then they change into the jagged, rocky summits of the Faroes rising above the clouds. It is journey's end for us. The airman is never safe from surprises and we scout around for a while. The clouds seem to be sliced open and the sun-bathed islands lie ahead. We make a sweep far out to the West to gain height. In a moment we have the big Lundefjord in sight; the little town of Thornshavn lies just behind a promontory. So far not a ship in sight. We dip into every little fjord to guard against shocks.

"I turn and fly into Thornshavn fjord. There is no sign of activity. Thornshavn itself must be just behind the rise. The bomb release and sights are set so that all we have to do is to dive. And now our real target comes in sight. A big merchant ship is lying at anchor across the fjord, with several smaller vessels behind it.

"We begin our attack. Everything goes off perfectly smoothly. The diving aircraft plummets down. We cannot miss. The ship gets bigger every second and we can pick out objects on the deck.

"Bombs gone! I have not even noticed my thumb pressing the little red button. While I am bringing up the nose of the aircraft I hear the voice of my gunner on the intercom: 'Bombs away!' A few seconds later: 'Hit on port side, first bomb too short.' We sweep round to inspect our handiwork. The poor devil must have collected a packet as dense clouds of black smoke are pouring from the deck and sides. But we must not linger. We have to get home and there is a long run before us.

"I glance at the clock. It is 16.04 hours on Sunday afternoon. We cannot forgo the pleasure of a victory parade above Thornshavn lying peacefully below. We take a good look at the anti-aircraft positions round the harbour, but not a shot comes our way. The gun crews are obviously having their afternoon nap. Before turning for home we fire burst after burst from our heavy machine-guns into the gun positions.

"Before long the Faroes are far behind us. The ship is naturally the main subject of conversation. It had certainly been heavily damaged, but was it still afloat? We send our report to base: 'Have attacked 7,000 to 8,000-ton ship. Successful.' But we are not yet safe home. Many an unpleasant surprise may be in store. We keep mighty close to the cloud base so that we can dive into it if an enemy aircraft turns up. To our right the northern tip of the Shetlands looms up, a spectral form. We give it a wide berth and set course for our home port. Heinrich Thies, my navigator since before the war, has his hands full. The aerials never get a minute's peace. We report our approach. Before long the familiar silhouette of the Norwegian coast emerges and a few minutes later we are down.

"I make my report and the same day the morning's sortie is the subject of a conference. We are all burning to know the fate of the ship. Early next morning two aircraft set out on an armed reconnaissance to the Faroes. We are in luck, as the weather, ever unstable in those latitudes, has held. One of the crews reports that our ship is half under water, burnt out and listing badly. Photographs of the bomb hits make it quite clear that her tonnage is at least 10,000. On its return this crew manages to shoot down an attacking Blenheim off the Shetlands.

"Next evening we are off again. We are conducting a blockade of England."

"The war against the British-American convoys to Murmansk and Archangel which we have been waging for months in the teeth of an ever mounting defence is the biggest open water undertaking in which the Luftwaffe has been employed hitherto. The squadrons engaged have already seen service in the Mediterranean and the northern section of the North Sea, but in neither of these areas have operations on such a scale and over a distance comparable to that between the Norwegian coast and Spitzbergen been required and carried out as a matter of daily occurrence. To make matters worse, Norway is anything but an ideal base for long-range air operations. Today, north of the Arctic Circle, there is a whole network of landing-grounds which heavily laden bombers can use, and which provide everything that aircraft and their crews require for their operation. But though this achievement in the face of unexampled territorial and transport difficulties cannot be praised too highly, it does not mean that our squadrons need only know how to fly and drop bombs. Taking-off and landing on cement and wooden runways in narrow valleys enclosed by hills and ridges over a thousand metres high would normally be regarded as aerobatics or lunacy. From the point of view of morale also, the strain is far greater in northern waters than in the south, as even in the warmer months the North Sea offers a ditched crew little prospect of rescue. Under circumstances such as these a Luftwaffe with no more than an average standard of technique and training would have been unable to prove a serious menace to this particularly effective highway to the Soviet Union. The success of the Adler and Löwen squadrons in barring this highway for the critical months must be regarded as one of the most important achievements of the year 1942.

"After the co-operation between aircraft and submarines had given us some outstanding successes in isolated attacks in the winter and spring, and the British Admiralty had countered with additional escorts and stronger A.A. defences on merchant ships, our first great blow was delivered at the end of May. From the 25th to the 29th of that month over 100,000 tons of shipping was sunk close to the ice-barrier south of Jan Mayen and Bear Island in continuous attacks which were favoured by the midnight sun. Simultaneously the port installations of Murmansk and Yokonga were bombed and toll was taken of cargoes which had survived the sea voyage. It was six weeks before a fresh convoy was assembled in Iceland waters for the northern adventure. Taking advantage of the fact that the ice-barrier had receded far to the north, this convoy was able to keep so far away from the coast that it reached the Barents Sea without serious loss. But on the 5th July our bomber squadrons had their greatest day; within a few minutes 40,000 tons was sunk and a further 40,000 tons heavily damaged in a few minutes. A cruiser of the protecting squadron, stationed at some distance, was put out of action. The officers commanding the convoy then lost their nerve and gave the order to disperse. Singly or in small groups, the merchant vessels tried to make their way to the White Sea. They

were chased and ruthlessly attacked all the way to the coast of
Novaya Zemla. In a series of fierce encounters our bombers ac-
counted for 200,000 tons.

"It looked as if this catastrophe had cured the British of all taste
for such risky ventures. For eight weeks our reconnaissance aircraft
on their daily flights reported the same story—North Sea clear. But
in August our inspection of the harbours of Iceland revealed signs
that a fresh attempt was in contemplation. Stalin's incantations
were bringing another convoy to the assembly point at Reykjavik.
In due course the familiar performance was repeated. Battleships
and cruisers sailed on mysterious courses beyond our range designed
to scare off our heavy ships or entice them out, and then disappeared
somewhere between Iceland and Greenland or cruised back and
forth in positions from which interception would be easy.

"On the 12th September, the scout reported the convoy as
coming within range. What we had suspected duly materialized:
this time there was an aircraft carrier. Two other features left us in
no doubt that the encounter ahead would be a tougher proposition
than our previous sorties."

So much for the press accounts of our squadron's attacks on
the Anglo-American northern convoys. Let me add a few extracts
from my diary which may bring out the human side of these bold
ventures:

"Yesterday we flew again—first the torpedo aircraft and then we
dive bombers.

"The roar of engines, running all out, resounds all over the
aerodrome. Clouds are hanging low in the fjords. The mountains
have hidden their bald peaks in the mist as if averting their eyes
from the murderous activities of these humans who have fallen out
of a clear blue sky on to this little bit of paradise. But the determination and audacity of these pioneers of the air were not to be denied.
In a few weeks they made a broad landing-ground among the cliffs
and boulders. Mountains were carted away, valleys filled up, roads
built: dredges, bulldozers, blasting. Expert labour units worked day
and night. And now we are taking the air, with torpedoes and the
heaviest bombs.

"The PQ, the convoy, is reported on the route Reykjavik–
Bear Island–Barents Sea. It is hurrying across the North Sea with
its cargo of war material, tanks and aircraft. Its destination is
Archangel and its zigzag course our target.

"Banak, Europe's most northerly flying base, with its small
wooden landing-ground, which is little better than a rough cause-
way, was in recent months the point from which our sorties were
made. It may in fact be said that flying from morning to night in

any sort of weather, particularly when the North Sea is shrouded in snow and ice, is the airman's highwater-mark of hardihood.

"It is evening before we get back. Two of us are missing, Flechner and Eckardt. Are we already so hardened that we merely register the news? Or is it that we feel that they have simply gone on ahead? Ought we not to envy them? But we cling desperately to life, a life that unrolls before our eyes as if it was but a day."

The entry of the U.S.A. into the war—and its previous material help to the Allies under the Lease-Lend agreement—had definitely shaken Germany's position by the end of 1942. Though the land fronts were still holding, it was gradually but inevitably hastening the collapse of German resistance. In the quickly built Victory and Liberty ships an absolutely inexhaustible supply of war gear and strategical raw materials was brought across the seas to the fighting fronts, and this gradually secured the Allies an overwhelming preponderance, the ultimate effect of which was either ignored or grossly under-estimated by the German leaders. There was constant reference to the figures of sinkings, but the fact that these no longer kept pace with the enemy's new construction—much less exceeded it—was overlooked. There was far too great a tendency to facile optimism and to view developments in the most favourable light, instead of facing the worst and taking prompt steps to avert it.

Even by the beginning of 1943 the German submarine and air attacks on enemy shipping were markedly diminishing. The figures of sinkings in the Wehrmacht *communiqués* were dropping, whilst our losses—of which nothing was said—soon became out of all proportion to the results obtained. This was due not only to the continuously increasing and well organized enemy fighter and anti-aircraft defence but a direct consequence of their employment of radar in the war at sea as well as elsewhere. In Admiral Dönitz's report of the 8th July to Hitler he wrote:

"The course of the war to date has made it clear that in the operations at sea the Luftwaffe is destined to play an outstanding part. While control of the open sea is now unthinkable without simultaneous control of the air, the Luftwaffe has become of decisive importance in the fight in coastal waters.

"Yet most serious of all has been the failure of the naval air force in connection with the navy's critical task, the submarine war. As the greatest difficulty facing our U-boats has been not in attacking their targets but finding them, there can be no doubt that wide-ranging reconnaissance from the air could increase the success of the submarine operations many times over. The influence of the enemy's air force in fighting the U-boats has been very marked; in fact it can be said that the crisis in the submarine war is the result of the enemy's control of the air over the Atlantic."

This pronouncement by the head of the German navy underlines the sins of omission in the building up and development of the Luftwaffe, which was not equal to the task assigned to it in the strategic operations in the Atlantic because it had not been system- atically prepared for that task. German air-sea warfare was a system of expedients, expedients imposed by the enemy. The German High Command was always trying to answer the *fait accompli* with inadequate and improvised means and methods. The really old- fashioned kind of naval warfare conducted by both the British and the Americans, with its innate clumsiness in the planning and execution of major operations, should have enabled us to recognize them promptly and adopt countermeasures. But appropriate offensive weapons were neglected as well as the uninterrupted development of an aircraft suitable for long-distance strategic reconnaissance at sea. This applies particularly to the German air torpedo.

Even before the First World War experiments to launch torpedoes from aircraft had been made in Italy and England. All experts regarded the torpedo-carrying aircraft as a promising military weapon. On the 12th August, 1915, Flight Commander C. H. K. Edmonds of the R.N.A.S., flying in a Short Seaplane Type 184, managed to destroy an enemy supply ship in the Sea of Marmara with a direct hit from a torpedo.

Experiments with airborne torpedoes had begun in Germany a little earlier and in 1916 the first torpedo squadron was formed in Flensburg. Up to 1918, 36,000 tons of merchant shipping had been destroyed and a Russian destroyer sunk by this means.

After the war all experiments with airborne torpedoes came to an end for a time in Germany so that the Naval High Command had to buy the Norwegian "Horten" aerial torpedo in 1933 and in 1938 the torpedo developed by the Whitehead concern in Fiume. But the military experts were still doubtful about the practicability of this weapon and did little before the war to develop a suitable torpedo plane, or even equip existing aircraft types with effective aiming and release gear. It was not until 1940 that those who regarded the airborne torpedo as a valuable weapon in the war at sea could make themselves heard. Yet even then the improved Horten torpedo, the LT F 5, was the only torpedo available to the Luftwaffe.

The German types adapted for using torpedoes could carry two of them, under the fuselage or the wings. They were the Heinkel He 115 seaplane and the He 111, Ju 88, Fw 200 and Do 217 land planes. The He 177 bomber was intended to carry up to six torpedoes. Up to 1942 the development and production of airborne torpedoes was the concern of the navy.

While in England, Italy and Japan, torpedoes with delayed-action fuses were in use by 1940–41 the German navy had no such fuses even for their naval torpedoes, much less for air torpedoes. The Luftwaffe was thus compelled to introduce an Italian fuse, which generally speaking, revealed nothing but defects in tests.

In view of the Japanese success at Pearl Harbor there was renewed agitation for the employment of the airborne torpedo.

Its production was faced with great difficulties owing to the high consumption of metals involved. The LT F 5, for instance, required 112½ lb. of chromium, 250 lb. of aluminium, 172 lb. of lead, 121 lb. tin and 825 lb. of copper. It was shortage of these raw materials and not of manpower, machine tools or factory space which impeded production.

While the navy output of torpedoes up to 1943 was only 800 a year and in that year rose to 1,200 (in three factories) the Luftwaffe managed to get the figure up to 3,000. In 1943, following a decree of the Armament and War Production Ministry, development and production were stopped once more, so that some very promising German developments—such as the bomb torpedo, which, produced in various calibres, made high-level fighter-bomber attack on ships possible—could not be exploited at a critical moment. The bomb torpedo was mainly intended for use against invasion ships. When the latter appeared at very close range off Sicily and subsequently off Normandy, neither suitable aircraft nor bomb torpedoes were available to deal with them.

Among the most important German inventions and developments during the last war were guided missiles, generally known as FK. The two most remarkable types of this new weapon were the P.C. 1400 FX (also known as Fritz X) radio-controlled missile, invented by Dr. Kramer, and the Henschel Hs 293 glider-bomb, an invention of Professor Herbert Wagner. Both were ready for use in war by the spring of 1943.

After the battles with the northern convoys in the summer of 1942 action against ships had practically ceased. Dive-bombing and low-flying torpedo attack on the last convoys (PQ 16 and PQ 17) had resulted in rising losses after the fighters from the escorting carriers and the enormously improved and effective anti-aircraft fire from the ships proved themselves better able to cope with our tactics. There was no escape for the Ju 88 when it was really caught in the new defences. Our methods became out of date, both from the tactical and technical angle. At this moment science was obliging enough to offer the Luftwaffe an entirely new weapon of incalculable value—the guided missile—for the air-sea war.

The characteristics of the two outstanding German inventions were as follows:

The P.C. 1400 FX missile had approximately the effect of a 3,000-lb. armour-piercing bomb. It was specifically designed for use against heavily armoured targets such as warships. It contained relatively little explosive but had exceptional penetrative power. Its tactical employment called for certain conditions.

In attack the target must be approached as in level flight with normal bombs. The most favourable altitude is above 15,000 feet. The observer aims with his Lotse bomb-sights and releases the FX. Its course can be directed by the guiding apparatus. If the observer sees that the FX will fall short, a movement of a little lever will affect the falling missile correspondingly and lengthen its course. The method of attack requires that the target, however high up, must be flown over and that the last phase of the flight should be in a straight line to permit accurate aiming. This is, of course, an unfavourable factor. Good weather conditions are also necessary, e.g. with the underside of cloudbanks at not less than 10,000–15,000 feet.

Despite these hampering conditions, this method of attack was a material improvement on normal bombing, as it was easier to hit moving targets. On the 14th September, 1943, the Italian battleship *Roma* was the victim of an FX attack south of Sardinia when it was escaping to join the Allies.

The Hs 293 was approximately as effective as the SC 1,000-lb. bomb which had already been employed with great success against merchant ships, lightly armoured warships and similar targets. As soon as it was released it gained speed under rocket thrust, and went ahead of the parent plane. The observer used a control lever to steer it up to the moment of impact. It was possible to employ this flying bomb in almost any tactical situation. The attack could be made from a distance of eight to nine miles from the target, i.e. out of range of anti-aircraft fire from the ship attacked. The Hs 293 was powerful enough to sink merchant ships of considerable size. Yet the timely employment of this new weapon was at first forbidden by the Luftwaffe High Command as it was said that Hitler would not release it for security reasons. This security business was carried so far that even the commanders in the field could not be told of the new weapon, even though stocks were accumulated and prepared for subsequent use there.

As General of the Bombers, I was in charge of the guided missile development and in that capacity I laid before the Air Staff proposals for attacking the Russian warships lying off Leningrad

and Kronstadt, the Black Sea fleet and the North Sea convoys. As
late as March, 1945, the Air Staff, appealing to Hitler's security
decision, objected to using the last Hs 293s against the Oder
bridges, though the few occasions on which they had been employed
clearly showed that the Germans had succeeded from the start in
the long-range guiding of missiles and bombs.

Operations on the grand scale with guided missiles were hampered by a shortage of suitable aircraft. The selected bomber, the
He 177, never appeared in quantity at the front. Previously, some of
the test planes of the He 111 type had been used and lost in carrying
supplies to Stalingrad. So in 1943, as a last resort, the last Do 217
aircraft, the serial production of which had ended, were transformed
to equip an FX group. But their range was so limited that it was no
longer possible to use them effectively in the Battle of the Atlantic.
The enemy convoys simply kept out of range. But in the Mediterranean, off the Portuguese coast and during the invasion the great
possibilities of this weapon were clearly revealed on the few occasions
it was employed.

At a demonstration of the V-1 and guided missiles at Peenemünde
Admiral Dönitz drew me aside and asked me whether the ship
which had been the target for some guided bombs had really been
hit. When I assured him of the fact he invited me to a conference in
Berlin. On this occasion the critical position at sea was very frankly
discussed. It was August, 1943. After I had presented my proposals
Dönitz simply remarked: "Why don't you say all that to your own
C.-in-C?" When I replied that for the time being I was unable to
get in touch with him as his personal staff were keeping me away, the
head of the navy remarked with a sigh of resignation: "Where's the
Reichsmarschall stag-hunting now? I haven't seen him for weeks."

Subsequently I saw a lot of Dönitz, who, strange to say, was kept
out of all discussions about the conduct of strategic air-sea war. But
at that time he had not yet got Hitler's ear on the fundamental
problems of the conduct of the war. He got it eventually, and even
persuaded Hitler to appoint the last Chief of Staff of the Luftwaffe,
General Koller. But by that time our air force had been destroyed
on the ground.

The wildest confusion prevailed in the production of the guided
missile. In the spring of 1943 the output of He 293s was to be raised
from 300 to 950 a month within a few months. Hertel, the Engineer
General responsible, reported that from April onwards 750 Hs 293
and 750 FXs would be produced monthly and an increase to 1,200
was possible. At the same time, Staff Engineer Bree, the technical
expert in the manufacture of guided missiles, told us that production,
especially in the case of the FX, was in a very bad way, mainly

because of a lack of factory space. We had a fully developed weapon but could not produce it in quantity and employ it. When Sauer took over the Fighter Staff one of his first steps was to order the immediate transfer to fighter construction of the technical personnel engaged in guided missile production. That was the end of the new weapon.

Once again—and it is significant of the internal stresses at the time—the technical experts battled with the stupidity and short-sightedness of the military and political leaders to secure the necessary recognition and exploitation in the national interest of the weapon they had brought into being. In a report dated the 15th August, 1944, a group of Rechlin engineers tried to open the eyes of Reichsführer Heinrich Himmler, then the most influential person, to the importance of this new weapon, and the fact that it would perhaps decide the war:

"The enormous importance of guided bombs to the course of the war is in no way realized, even today. The results to date—forty per cent of the bombs discharged at the enemy have been direct hits, though the weapon is quite new and the launching conditions were very difficult—are not known to the men whose decision alone matters. The order of the Reichsmarschall that guided missile production must stop at once was carried out in such a way that the bombs, eighty per cent of which were ready for use, were condemned to destruction, and aircraft completely adapted for launching them were modified once more for reprisal attacks on London, so that it is no longer possible to give these bombs, even for training purposes, to the units familiar with the weapon."

It was too late. Most of the guided missiles had already been broken or blown up.

Despite the poor opinion of their weapon in high quarters, the inventors had gone on improving it. To prevent enemy wireless interference the radio receiver was removed and successful experiments were made to steer the falling Hs 293 by two trailing wires attached to it. A radar finder and a telescope eye were provided. Guided missiles of smaller dimensions for use against the Allied bombers were being got ready.

German scientists and technicians had shown by their revolutionary inventions that they could enable us to hold our own with the enemy coalition, despite its material superiority. If final victory could no longer reward these products of German brains, the short-sightedness of our leaders was primarily responsible.

In his interrogation after the war Colonel (G.S.) Bernd von Brauchitsch, Göring's senior Adjutant, remarked: "The Luftwaffe knew all about its own possibilities—how effective bombing attacks

7

could be carried out, how to deal decisively with ship targets, how the defence of the Reich could be assured and problems of transport mastered. All it lacked was the material means." The only proper answer was given him by General Koller, the last Luftwaffe Chief of Staff, in his comment on this statement: "Hence the failure to produce suitable aircraft and the decision to cease production of guided bombs?"

Göring's court camarilla was largely responsible for keeping him in the dark about guided missiles and other promising weapons and giving him belated or false information. Incredible as it may seem, Göring did not know that the Italian battleship *Roma* was sunk by us with an FX radio-controlled bomb. He was not convinced until I produced a film of the sinking at a subsequent conference. But even this revelation produced no change of heart.

In the year 1944 enemy sea traffic was again proceeding practically unhindered in the Atlantic and round the British Isles. Every month hundreds of thousands of tons of American war material were landed in British ports and England was converted into a strategic aircraft carrier and troop transport from which the decisive jump for the re-conquest of the Continent would be made.

To the German Reich, already struggling desperately in the East against ever mounting Russian superiority, this defeat in the Battle of the Atlantic meant a "Second Front" in the West. The issue of the war was no longer in doubt. Despite our lightning victories and vast successes on the Continent at the start, the war had already been lost in the depths of the Atlantic and the steppes of the Ukraine before the surge into Germany, the hard core of the real resistance, began.

CHAPTER IX

DIVERSION TO CRETE

AFTER the military occupation and political incorporation of all the Balkan States the Reich found itself the Soviet Union's next-door neighbour from Kirkenes to the mouth of the Danube. It was separated north, west and south, by the sea from its British opponent. Successful prosecution of the war against the latter called for the production of a combined sea and air strategic design by the German leaders, yet Hitler, with his Continental outlook, kept his eyes fixed on the rosy east, which he had already indicated as the German people's future *lebensraum* in his book *Mein Kampf*.

The previous course of the war with England had also shown that German sea and air power was inadequate to cut off the British motherland from her oversea sources of supply and force her to her knees in isolation. British toughness had triumphed over the "blitz" and London was even less inclined to negotiate than before.

It was plain for all to see that the time had not yet come for Hitler's own political plan for carrying England with him against Russia. Moscow, on the other hand, had no intention whatever of intervening actively on Germany's side in the war with England. In Stalin's words, the Russians declined to "pull the chestnuts out of the fire for the imperialist capitalists", and preferred to be the *tertius gaudiens* at the end of the war. That policy had already proved its value in Poland, Finland and the Baltic States. Molotov's demands on Berlin in August, 1940, unsuccessful though they were, left no doubt about the Kremlin's further claims. Faithfully pursuing the traditional expansionist programme of Tsarist Russia, Moscow once again professed to regard the question of passage through the Dardanelles as very pressing—all the more because the Reich had become an unwelcome rival in this Russian sphere of interest after the victorious conclusion of the Balkan campaign.

The Führer's Headquarters was preoccupied with all these questions when the preparations for the greatest air-land operations were made at the beginning of May, 1941. The invasion of Crete was to be entrusted to the XIth Air Corps of Parachute and Air-borne Troops, commanded by General Student, supported by the flying squadrons of the VIIIth Air Corps under General von Richthofen and the Vth Mountain Division attached to the List Army Group.

General Student's plan was to drop his parachutists and glider troops on the British key positions with a view to engaging the garrison, mainly composed of New Zealanders, in immediate action. Meanwhile, the landing ships, comprising small naval vessels and improvised fishing boats, would carry out landings at various points on the coast in the enemy's rear. Several air transport squadrons, composed of over 500 of our tried and trusty Ju 52s and gliders, would keep the front supplied with reinforcements and material from the Greek mainland. General von Richthofen with over 640 Stukas, destroyers and medium bombers would concentrate his attacks first on the British naval units and then keep down enemy anti-aircraft fire during our landings from the air.

The island of Crete, a bulwark both for the Balkans and Asia Minor, was firmly held by the British who controlled all the approaches to the Greek harbours, the Dardanelles and the Straits of Otranto. It was an air base dominating the whole of the Balkan Peninsula. The Ploesti oilfields and traffic on the lower Danube were within range of British bombers. Crete also protected British positions in the Levant, on the Suez Canal and elsewhere in Egypt from attack by the Axis powers. The Italian Dodecanese were outflanked and rendered strategically impotent by the British fleet and the air bases on the island. The possession of Crete was of the first importance to England's position in the eastern Mediterranean.

Although the forthcoming German operation against Crete could not remain a secret to the British General Staff and the British fleet had complete command of the sea, the bold enterprise ended with a signal triumph for German arms.

But at the very outset that triumph was jeopardized by several seriously faulty dispositions by the German leaders. The weak German parachute units dropped straight into the enemy's concentrated anti-aircraft fire and were dispersed on landing and found themselves faced by opponents superior in number and material. The losses of our own air transport were considerable, over 150 Ju 52/3s being destroyed. It was only the arrival of the landing ships bringing in the mountain units that gave us Candia and Heracleion in the teeth of the fiercest opposition of the New Zealanders, who fought to the last man.

Our dive bombers and medium bombers carried out their tasks in exemplary fashion. Off the shores of Crete the British warships intervened and on the 21st May drove off the second wave of German landing ships which were bringing reinforcements from the island of Milos. But thanks to our activity in the air the British fleet did not dare to bring in any reliefs for the hard-pressed garrison on the island. The Luftwaffe had shown its superiority over the British Mediterranean fleet and once again it had been demonstrated that even strong naval forces are helpless against an opponent superior in the air unless it has destroyers and fighter cover.

As in Norway the previous year and also in the Balkan campaign just concluded, the R.A.F. was conspicuous by its absence in the struggle for Crete. Its strength at that time was adequate for the protection of the home country but not for that of the far-flung British Empire, with its spheres of interest and vulnerable outposts. With the possession of Crete fate had brought the German High Command as close as possible to what was perhaps the weakest and most sensitive point in the British world empire. The question was whether the Führer's H.Q. had the sense and the necessary subtlety of geopolitical instinct to follow where fate beckoned.

The fall of Crete opened up all sorts of new possibilities. In German hands it guarded the whole Balkan peninsula, controlled— in conjunction with the Dodecanese—the approaches to the Dardanelles and the whole southern coast of Turkey, a development which would have considerable effect on the Ankara Government's appreciation of the situation. Cyprus, the Levant coast, Alexandria and the Suez Canal, the Nile delta and Cairo were within German medium bomber range. Wavell's positions in North Africa were not only threatened by Rommel on the West but could easily be shaken by air attack from Crete. The Red Sea, the Arabian oilfields, Bahrein, the Anglo-Iranian Oil Company's refineries in Abadan, the Mosul fields with the pipe-line to Haifa, the Trans-Iranian railway and the whole of Turkey were at the mercy of our long-range bombers. From Crete the decisive blow to the British route to India and England's key position in the Middle East could be delivered. At that moment it was the centre from which the British Empire could be mortally wounded without the great risks attaching to air attack on the motherland itself.

The fall of Crete had contributed substantially to shake Great Britain's prestige in the Moslem world. Founded on force as it was, it was easy for a yet greater power to shatter British influence among the Moslem peoples once and for all. The defeat of France had given life to the nationalists in Syria and the Lebanon. Palestine had long been in turmoil. Under El Ghailani Irak was in open revolt, which was receiving feeble support from occasional German airborne transport. Ibn Saud was torn between enticing dividends from Standard Oil and a favourable chance of squaring accounts with his old rival, the pro-British Emir Abdullah of Transjordania. In Rome and Berlin Egypt was insisting upon its neutrality. England was rapidly losing face in the East and it would need no

very great effort by Germany to hasten the process at a moment so
favourable for us.

It was not until much later, towards the end of the war in fact,
that I was commissioned to render a vital service to the Grand
Mufti of Jerusalem, El Husseini. Several highly interesting conferences were held at Führer's H.Q. and Himmler's headquarters.
We were still in the dark as to the fate of the German garrisons in
Crete and Rhodes so that the choice of the south-eastern jumping-off
ground Vienna–Aspern was forced upon us. In those circumstances
the prospects of a successful operation seemed very speculative. Yet
a week after the start the big Ju 290 transports were home again
after carrying out their mission.

But in 1941 the indicators on the situation maps at Führer's
H.Q. pointed east, not south. Of course our propaganda did not
neglect Hitler's words about there being no more islands and it was
officially emphasized that Crete was to be regarded as a sort of dress
rehearsal for bigger operations on the same lines. But in fact the
plans for an invasion of England had to be laid aside as impracticable,
as meanwhile R.A.F. Fighter Command had become so strong and
well handled that in conjunction with the night fighters and the
elaborate radar system it deprived a German landing from the air
of any chance of success. The German navy was too weak even to
think of giving the necessary cover to the waterborne transport
that would be required, relatively narrow though the English
Channel is. Since 1940 and the days of Dunkirk the situation had
changed fundamentally.

Only weak German and Italian air squadrons were left on Crete
and the Dodecanese and their activities for the rest of the war were
confined to isolated operations of little or no importance, though it
is true that long-range German reconnaissance aircraft managed to
photograph the oil tanks at Haifa and the Egyptian Pyramids, keep
an eye on traffic through the Suez Canal and even chase the R.M.S.
Queen Mary out of the Red Sea. But we lacked the bombers for
effective attack on those enticing targets.

England gained the time she required to restore her threatened
position. Wilson brought the British 9th Army from Baghdad to the
Levant coast. In North Africa Montgomery succeeded in maintain-
ing his position and gathering together reinforcements and supplies
while Rommel's strength drained away in the sands as he used up
his last reserves. The Alexandria squadron ventured out into the
eastern Mediterranean again and British submarines, to be soon
followed by destroyers and fast launches, operated without let or
hindrance off the coasts of Crete. Supplies to our garrison there thus
became increasingly difficult and at times were limited to air lift.

After Badoglio's capitulation in Italy the main activity of the German forces on Crete was taking their former Italian allies prisoner and keeping the ever restless Cretans under control.

So Crete remained a sideshow without decisive effect on the course of the war, a military *pièce de résistance* which was denied a victorious conclusion by our leaders themselves.

CHAPTER X

TRAGEDY OF THE LONG-RANGE BOMBER

In the spring of 1941, after the first raids on Glasgow, we were summoned to a conference held by Reichsmarschall Göring in The Hague. I was also to report on our activities against ships and, above all, answer the question why we had failed to sink an aircraft carrier.

Towards the end of the meeting someone gave me a friendly tap on the shoulder. It was Colonel-General Udet, then Quartermaster-General (Air). He playfully remarked that what the Reichsmarschall was saying was not very important. Could not I collect a few friends and retire to a corner with him? He produced a bottle of brandy. As an old comrade of Göring's in the Richthofen "Circus" he could be free and easy in talking to us young officers and drink our health. In the course of the conversation he said that the Americans would soon be building their new bomber. It would probably have four engines. He had been in the States often enough to know what their industries could do when they were put on a war footing. Hitler and Göring wanted bombers, not fighters. That was the whole tragedy. We must keep his remarks to ourselves and if at any time we had any worries and wanted to get them off our chests we were to go and see him. That was Ernst Udet, a man who had won all the decorations and was the most successful surviving fighter pilot of the first war.

Udet's prophecy was to come true only too soon. A little later, in March, 1941, President Roosevelt commissioned the head of the Air Force, General Arnold, to build a strategic bomber fleet to defeat the Axis Powers.

What had Germany done in the way of producing long-range bombers? The fact is that after the death of Wever, the Chief of Staff in 1936, the plan for a strategic air fleet had been dropped because:

1. The service chiefs in principle assumed that there would be nothing more than a war in Europe. They thought that the European theatre alone would be involved.

2. They believed that in building up a modern air force such as Germany possessed just before the war they had done all that was necessary for the destruction of the enemies' air forces by a few

surprise blows at the outset of hostilities, so that nothing would be seen of them during the operations on land.

3. Germany's strained resources in materials, especially the important raw materials, made the production of a mass of big aircraft impossible.

So production of a medium bomber was regarded as the main target to be aimed at, particularly because it facilitated accurate aim in oblique or diving flight.

Despite the views of the Air Staff the Technical Department of the Air Ministry, in conjunction with our aircraft industries, had carried out considerable research into the problem of the four-engined bomber. There was the He 119, an aircraft with two engines side by side operating one airscrew which could serve as prototype for a four-engined aircraft with two airscrews. A special four-engined aircraft from the Heinkel works, the He 116, was able to set up several international records in 1939. Towards the end of the same year the experiments had proceeded so far that from the two prototype He 118s and the He 119 an aircraft was to be developed which could be fitted with four engines, two coupled engines operating one airscrew mounted on each wing. The new prototype having given satisfactory results Heinkel was told to go ahead with production and lose no time, because the Luftwaffe possessed no useful aircraft for operations over water, especially the Atlantic.

Notwithstanding the compromise with Chamberlain reached at Munich and Godesberg, precautionary measures for a possible conflict with England were put in hand. Professor Ernst Heinkel developed his He 177.

The coupled-engine arrangement of the He 119 was not successful in the He 177 and from the first day broken pistons, liability to catch fire, an inadequate cooling system and other technical defects led to continuous delays in mass production. The requirements for the He 177 were a speed of 335 m.p.h. and a range of 3,750 miles, as well as perfect ability to dive. It was this last requirement which worried the industry most and was the final reason why the aircraft was equipped with coupled engines. The requirements could never have been met by an aircraft with four engines and four airscrews.

At the subsequent conference at which Göring bitterly complained that the He 177 had not yet put in an appearance though it should have gone into service after 1940, he said to Professor Heinkel:

"It's idiotic to ask a four-engined aircraft to dive. Four-engined aircraft don't need to dive. I was glad enough when a two-engined plane managed it. Well, there we are!" "The airframe must be strengthened for diving," replied Heinkel.

Göring: "It need not have to dive!"

Heinkel: "In that case it can go into immediate service."

Up to September, 1942, of 102 He 177s produced the Quarter-master-General had received only thirty-three. Immediately after this aircraft was delivered to the service, Hitler insisted on its immediate employment, even in the simplest form, in the Russian theatre. By "simplest" form Hitler meant "level flight at night against ground targets so far away that they could not be reached with other aircraft".

Göring hastened to remind Heinkel of Hitler's wish and give his own idea of the way in which the new aircraft should be flown. When Heinkel carelessly replied that in that case his He 177 was "ready now" Göring told him that his tactical requirements were as follows:

"Your He 177 must be able to carry torpedoes externally. Diving isn't required for that. Not even dipping. The nearer it is to level flight, the better. It must also be able to drop special bombs on ships from far away. If I want to go to Sverdlovsk or somewhere like that it must be able to fly pretty high. But whatever you do, give me something that is a real long-range bomber, can fly very far, carry a decent load, and above all reliable and safe. The engines are of course the main thing, so that it can range far over the sea and attack convoys."

Göring's insistence was natural. It was inspired by the situation, which was getting more critical on all fronts. His bitter reproaches to the industry were not entirely justifiable, however, as his ban on experimentation had left new weapons and aircraft—including the He 177—bringing up the rear. The real reason for that step was the overwhelming initial victories of the Luftwaffe, which in the eyes of G.H.Q. made the production of a big bomber seem a minor matter. At that time it was believed that the requirements of the air war could be met by mass production of the twin-engined bomber that had proved itself in the blitz.

At the end of 1942, of the He 177s which Hitler had demanded for his campaign in the East one squadron was employed at the end of 1942 in supplying Stalingrad from the air. Up to the 29th January, 1943, four aircraft had been lost through the engines catching fire (though Heinkel had said in September that the fire risk was "as good as cured") and the rest held up by delays.

Thereupon Milch was hauled over the coals by Hitler, who had again taken up the He 177 question. Milch told me about this meeting on the 12th March, 1943:

"We are in hot water because the He 177 is not all right. The Führer has been talking to me. I felt like a little schoolboy who hasn't learned his lesson. I tried to explain things to him. But it is difficult to explain things to someone who does not follow our train of thought particularly when one cannot simply say: 'I've had to take things as I find them; I can't help it; it's the responsibility of my predecessor.' "

Milch's recourse to Hitler's technical ignorance and his mania to throw the blame on Udet, who had been dead for eighteen months, was all talk. I knew myself from other occasions when I met Hitler at technical conferences that the truth was quite otherwise. Intuitively, Hitler often had the right idea as to what was technically possible or not. In the case of the He 177 he had expressly said that he considered the engine arrangement unfortunate. When I visited him in March, 1945, with Colonel Knemeyer to discuss a number of thoroughly complicated technical processes we had a very lively talk such as was only possible among the initiated. Hitler wound up with the words: "Why haven't we met before. Thank you, and I hope we shall often have further talks."

On the 30th March, 1943, Göring was vocal again: "I was promised a big bomber, the He 177, which should have been in service a year ago. But when it was tried out, it had catastrophic losses, and not in action either. A year has gone by, and when some sort of a thing comes along in a year or so it will probably prove to be hopelessly out of date."

It had already become out of date. As a result of the eternal modifications its performance had fallen so low that there were many objections to its employment as a long-range aircraft in the jobs for which it was intended.

In June, 1943, Grand Admiral Dönitz made a report to Hitler: "The declining figures of sinkings in the U-boat war can only be made good by making more use of the Luftwaffe." On an order at top level, the He 177 was immediately put into service as escort for our submarines far out into the Atlantic. Though Milch knew well enough that the outstanding defect of this aircraft, the coupled engines, had not been remedied, he reported that the He 177 could be employed at any time for high-level attack and would be equally useful for submarine escort duties. This report was in direct conflict with the views of the "Fliegerführer Atlantic"[1] based on actual experience, of the usefulness of the He 177 for the sea war: "With a tactical range of only 930 miles the He 177 certainly cannot be used in all areas open to long-range reconnaissance. Its useful range does not extend beyond the area west and north-west of the Bay of Biscay." Thereupon the Air Force Staff issued an order: "In the light of first experience with the He 177, modifications of this aircraft must be taken in hand at once. The value of this class for

[1] The officer directing air operations in the Atlantic zone. (Tr.)

dealing with ships in distant waters will be greatly enhanced by the use of guided missiles. The most urgent task is to extend its range beyond the existing limit."

The German aircraft industry was no longer in a position to carry out that order. Quite the reverse. Further modification of the He 177 reduced its range and made it unsuitable for the mounting demands of the Battle of the Atlantic.

Meanwhile, the Junkers concern had developed its new long-range bomber, the Ju 288. This was faster than the He 177 but its range was less and it had the same defect of the coupled-engine layout. So both aircraft were dropped from the sea war and were now offered to Major-General Pelz for the resumption of the "retribution" campaign against the British Isles. The General voiced his natural objections: "I must emphasize that the He 177 is 100 m.p.h. slower than the Ju 288."

As it had been decided to continue with mass production of the He 177, work on the Ju 288 was stopped. Although Pelz always fought against the He 177 and preferred a fast medium-range bomber like the Ju 188 for the operations against England, he had to yield to "higher" views. He therefore ordered that five to seven bomber groups should be equipped with the He 177 for use against the British Isles with at any rate *some* prospect of success. This would have involved a replacement of 200 a month. At that time the production of fighters was already far more urgent as the result of intensified air attack on the Reich. Five hundred fighters could be produced with the same expenditure of material and labour as 200 He 177s. In addition, the enemy raids had made the fuel position so critical that on that ground alone the employment of big bombers was no longer feasible.

On the 3rd July, 1944, Saur was able to persuade Göring to order the production of the newly modified He 177B with four separate engines. When Hitler approved this decision he added that all the old He 177s should be scrapped. And so the most tragic chapter in the history of German air armament came to a close.

It would leave the story incomplete if I did not make mention of some other projects which were intended to lead to the production of a long-range bomber.

From the outbreak of hostilities a small group of designers had been active in producing aircraft with sufficient range to cross the Atlantic and reach America. It was the Quartermaster-General who took the initiative; the Air Staff had expressed no views and given no instructions. The industry was asked to provide a range of at least 7,600 miles; i.e. the distance from Brest to New York and back. Allowing for the necessary reserves, that meant a range of 9,300 miles. An anticipated load of three to five tons of bombs plus the usual armament and other equipment were to be provided for.

On this specification, the Focke Wulf and Junkers did research work producing on paper aircraft of one hundred to one hundred and forty tons flying weight. But as no sufficiently powerful engines existed, final production would have taken three or four years. Focke Wulf also suggested an "emergency" long-range aircraft of wood, the Fw 191, with two or four engines and simple defensive armament. But the works had no free production capacity and in any event wood was not approved.

On his Me 261 (looking like a sealed-up Me 110) Messerschmitt based his four-engined Me 264 design, which was offered to Göring in the middle of 1941. Fresh requirements were laid down and this aircraft's performance became adequate for the Battle of the Atlantic. Flights to America were planned, the extension of range being obtained by refuelling in the night from a tanker aircraft accompanying the first for 1,850 miles. Experiments in refuelling in the air were in progress before the war.

In May, 1942, new designs were offered by Heinkel, Focke Wulf, Junkers and Messerschmitt. The Junkers was the six-engined Ju 390. The Messerschmitt specification was: weight 50 tons, range 8,000 miles, useful load 3 tons, armament and armour 3 tons, four DB 603 engines. It retained the old designation of Me 264.

The Technical Department decided on urgent production of the Me 264. It was to be ready for service at the beginning of 1944. But here again the "Messerschmitt bogy" was in existence. There was one hitch after another. The thirty-hour test of the DB 603 engine had to wait until 1944. Refuelling from the air at night (necessary for the American trip) required further experimentation in the case of the Me 264.

On the 12th May, 1942, Milch had put to Jeschonnek, the Chief of Staff, the vital question: "Don't the General Staff really believe that by refuelling from the air at night we can send a bomber vast distances, probably several thousand kilometres from its base?" "There's no object in doing so!" was Jeschonnek's quick and peremptory reply.

When I urged, as I frequently did, that such refuelling was the solution of all the difficulties surrounding the American project, Jeschonnek used to say: "I agree with you—but there's Russia!"

It was no use thinking of even the modest operation against the American continent while Russia was devouring all our available resources in aircraft.

Göring, who had placed great hopes on a long-range bomber, gave vent to his disappointment in March, 1943: "I well remember that at Augsburg—it was exactly a year ago—I was shown an aircraft that really called for nothing more than to be put into mass production. It was to fly to the east coast of America and back, from the Azores to the American west coast, and also carry a lot of bombs. I was told so in all seriousness. But in those days I was still so trusting that I half believed it."

The great dimensions of the aircraft were an obstacle to its employment over water. At first normal bombs and torpedoes were provided. But guided missiles seemed the most suitable. Experiments were in progress to attach little aircraft which could be employed as fighters or to make bomb attacks on ships. After the attack they could be made to return and be re-attached to the mother ship.

June, 1944, was the last time the production of "extreme-range" bombers was discussed. Saur had this to say on the subject: "to talk with the Führer shows me that his real view is that we must draw prompt and firm conclusions from the decision to concentrate in aircraft manufacture. Now we know that the right line is to produce revolutionary inventions and leave botches and bungles alone, I am in favour of going all out with the Me 264. This Me 264, with its combination of piston engine and turbo-jets, offers us prospects of revolutionary importance. Its load and range make it seem adapted for all purposes, even in relatively small numbers." (Messerschmitt had offered his Me 264 with additional turbo-jet units and promised that he could soon produce three or four prototypes in a branch of his works.)

After the Anglo-American invasion of North Africa I had proposed that we should inaugurate small strategic raids, which would have great effect, on North America while we were still in possession of advanced bases in Normandy and on the French Atlantic coast. I developed that idea at a private meeting with Jeschonnek.

In my opinion the most vulnerable point in the American system of defences is the Panama Canal, a man-made technical masterpiece with a highly sensitive mechanism which can be thrown out of gear by the least shock. Our U-boats were then operating in the Caribbean and it seemed to me more profitable to stage a project in conjunction with them rather than that they should confine themselves to bombarding paradisical islands which had once been the haunt of pirates or torpedoing the pretty ships of the *Flota blanca*. My plan, which was crammed with all the necessary naval and technical calculations before it was put before the Air Staff, was in the first place to bomb the Canal from some existing heavy transport aircraft which were already equipped for bombing. On the return flight these aircraft should surface close to our own submarines so that the crews could be taken off. I promised myself full success from such a bolt from the blue and anticipated far less risk than we should have to run every day in the Mediterranean or in the resumed "retribution" raids on England.

Secondly I thought—we had meanwhile learned of the American strategic bombing plan which was devised at Casablanca in January, 1943—it would certainly have a great moral effect if we carried out a "retribution" raid on New York. In this connection I made another proposal which was feasible with the aircraft and crews then available. It was a surprise attack on the American continent which we thought would create real consternation among the civil population and provoke panic measures in the way of air-raid precautions, air defence and so forth. A few flights would do great harm to the American economic and industrial programme as large quantities of personnel and material were held up in the U.S.A. For hours Jeschonnek forgot his worries about the bogged-down air war in the East, where Göring's fantastic project of supplying Stalingrad from the air was ever before his eyes. He sat up with me discussing all the details of my scheme until long after midnight.

On the following morning a considerable number of us were lunching with Göring. I was late and missed the chance of a prior talk with Jeschonnek. So I was somewhat surprised when Göring, after a few spoonfuls of soup, produced as his latest trump the American project, the attack on New York, on which he was keener than on the Panama Canal. I turned quite red, while Jeschonnek could hardly keep himself from laughing and gave me a knowing wink. Göring in his enthusiasm noticed nothing.

As usual, this was soon forgotten, as Göring would not co-operate with the navy and the idea of refuelling from the air at night was not seriously taken in hand. Once again, we could not see the wood for the trees. When the invasion in the West succeeded, all these long-range projects became illusory.

Germany's raw material shortage, on top of the claims of the other theatres, would have prevented us from employing large formations against America. The capacity of our industries was based on a continental European war and would have been inadequate for this task. A strategic bomber fleet could only be built up if the Russian campaign ended quickly. Such a fleet would have been dangerous to England and the Atlantic convoys, but not to America. At that time Germany possessed no weapon of annihilation, such as the atom bomb, with which to conduct that strategic air warfare on the intercontinental scale which a promising air offensive against North America would have called for.

CHAPTER XI

BEFORE THE STORM

Hotel Adlon,
Berlin.
2nd October, 1940.

. . . . BACK from my trip. The first part of my official report of the journey lies before me. Continuation superfluous. I wanted to make it without propaganda intent and have succeeded. *Honni soit qui mal y pense.* . . .

Travelling comparatively quickly and comfortably by the Berlin–Konigsberg–Riga Express, I was soon in Moscow the capital of the Soviet Union. Even on the way I was surprised by the patriotic tone of the Russian press. Is the Russian sphinx changing?

Shortly after my arrival the German Ambassador, Count von der Schulenberg[1] received me at his house. A sympathetic old gentleman who, according to his predecessor, Nadolny, is one of the best informed men on Russia and the Near East. Herr von der Schulenberg asked me for my impressions of my journey. I happened to be in Riga when the Soviet troops occupied it. Arrests and kidnappings in the open streets, communist processions and military parades had quite changed the face of this beautiful city in a few days.

In the afternoon the ambassador calls at the Kremlin while his amiable Counsellor, Gebhard von Walther, takes me for my first tour of the city. There seem to be neither rich nor poor. The streets are thronged with busy people. Old-fashioned isvostchiks compete with cars and trams crammed to bursting. One sees no beggars or idlers and smartness is unknown. No private shops. There are queues at state food distribution centres and the transport termini. Russian men, women and children wear cheap and shoddy clothes. No one smiles or looks happy. Life in Moscow seems one long hunt.

In the evening I sit with the Military Attaché, General Köstring, and my old friend the Air Attaché, Major-General Aschenbrenner, on the terrace of the very modern hotel Moskwa, not far from the Kremlin. The waiters are in evening dress and a jazz band plays swing. All the desires of a sophisticated human palate can be satisfied here if one has the necessary exchange. Many foreigners and diplomats are seen.

We have a long talk about the Russian Air Force and Russia's

---

[1] He was murdered after the 20th July, 1944, attempt on Hitler's life.

military strength. Both officers are greatly concerned. Their sober
reports go into waste-paper baskets or are watered down before
circulation when they reach Berlin. Aschenbrenner thinks that an
understanding with Russia is possible for quite a long time. Russia's
strength is in her system which is not going to collapse overnight. It
is quite impossible to find out the size of the Russian army. If there
are any tactical or strategic Air Force formations they are out of
date. The Russian soldier obeys Stalin's order blindly. The armament industries concentrate primarily on modern equipment for the
army and its auxiliaries. Russia cannot be conquered on land.

As we walked home across the well-lighted Red Square, it was
clear to me that these officers were wasting their time.

A car from the Intourist Bureau on Sverdlovsk Square took me
over poorly paved streets to the Severny Station, from which the
Trans-Siberian Express goes twice a week to eastern Asia via Perm,
Omsk, Tartarskaia, Novosibirsk, Irkutsk, Vladivostok or Manchuria
and Korea. The station square was packed. Red Guards in their
dark brown uniforms and carrying sidearms were stationed at the
exits. It needed no close inspection to discover that the Trans-
Siberian, the longest train in the world, is no blue train. Engines
and coaches had weathered many a buffeting and apparently
survived the World War and the ensuing revolution in fine style.

It was Wednesday afternoon when, to the accompaniment of good
wishes from many kind friends, I left Moscow. The train takes ten
days to Otpor, the Manchurian frontier station where the Asia
Express for Manchouli, Harbin, Mukden and Fusan was waiting
for me. It took a further five days via Shimonoseki to reach Tokyo,
my immediate destination.

Of the places through which we passed I well remember Sverdlov
(formerly Ekaterinburg, where the Romanov dynasty found its
bloody end) and the adjacent giant steel works, a wholly modern
city of iron and concrete with 200,000 inhabitants. In a very few
years it has risen from nothing. One of the greatest armament works
in the Soviet Union, at Stalin's command it has come into existence
on the frontier between Europe and Asia to be beyond the range of
enemy attack. But for how long? The iron industry of the Urals is
several hundred years old. Here was the main armaments foundry
of Peter the Great, associated with the names of the great firms of
Demidor and Stroganoff.

In the way of ores almost everything considered worth having is
to be found there—gold, silver, platinum, whole mountains of
magnetic iron, copper ore of the finest quality, lead, nickel, chrome,
iron, coal. Hence the name "Coal Alps". It is said that the deposits
must exceed fifteen hundred million tons, of which half has not been

opened up. Magnetic mountains exist containing three hundred million tons of ore with a 60 to 70 per cent iron content. There are few areas in the world in which such abundant resources of raw material offer such prospects for the establishment and growth of important industries. The Russian rulers know it. Industries shoot out of the ground like mushrooms and it is small wonder that today the foundations of Russian heavy industry extend from the Ural region to the Kusnietzky Basin.

Next morning, eight hours later, this watershed between two continents was behind us. When the train stopped for a few minutes at a little country station I noticed some apparently empty trucks with barred windows standing in a siding. They were a prison train used for the conveyance of convicts transported to Siberia. About a hundred miles south of Sverdlovsk, on the eastern slopes of the Urals, lies Cheliabinsk. It is famous for its triangular white marble pyramid, the "Monument of Tears". How much sorrow and misery has this pyramid seen as untold thousands of wretched exiles shambled past it, dragging their chains. With Russia, their homes and all that was dear behind them and nothing ahead but the timeless wilderness of Siberia, the desolate taiga, the gloomy forests and icy tundra ahead, they were not men but numbers, their lives not worth a bullet.

Banishment to Siberia was introduced in the 16th century. The first case was in 1591, when Dimitry, the murderer of the Tsarevich, and his accomplices were sent there. In 1823 the Transportation Bureau was set up in Tobolsk. Between 1823 and 1900, seven hundred thousand prisoners, followed by 216,000 members of their families, passed through Cheliabinsk. Among them were many who afterwards became famous—Lenin, Trotsky, Madame Nadezhda Krupskaya and Stalin. Was it surprising that the proletariat turned against their imperial oppressors after nihilism had been quietly flourishing in their midst for many years?

But that is only one side of the coin. My train rolled through splendid forests and meadows, where cattle, sheep and horses were grazing. The great majority of the twelve million population came voluntarily to Siberia, which they regarded as the Land of Promise, the land of future and fortune. They were attracted by the limitless spaces of this new region, the fabulous riches of its great mountains.

The story of its colonization is extraordinarily fascinating and instructive. No such mighty empire has been conquered at such trifling cost as Siberia in less than a hundred years. Seldom has so great a development in history been so much disregarded by the world outside.

I was left in no doubt that inside Russia itself the Marxist dogma of world brotherhood had been replaced by genuine patriotic fervour. As a realist, Stalin had soon seen that to carry out the Five Year Plan would involve great sacrifices, sacrifices of which only a nation inspired by love of country would be capable. That may be the reason why Lev Trotsky, the doughtiest champion of world revolution, trod the road to banishment in Siberia a second time (1927–28).

It was always the same with professional revolutionaries. Meanwhile, under Stalin's leadership, the youth of Russia were won over to a new point of view. Hundreds of thousands of young communists passed through the universities every year. They filled the key positions and their enthusiastic, patriotic work is an important contribution to the success of the Government's programme. They get higher salaries, better accommodation and other favours. So they are slowly turning into a new, privileged class.

The long-range target of Soviet policy remains the same—world revolution. Yet foreign policy is not mixed with domestic policy in Russia. And Russia has *time*. Lenin was the spiritual forerunner but Stalin is creating the solid basis. To him all means to that end are justified.

On the fourth day we traversed the Baraba desert. Our next stop was Novosibirsk, on the foaming river Ob. To the south stretch the still unexplored Altai Mountains, the greatest still untouched reserve of raw materials on earth. In the Kuznetzky Basin there are coal deposits estimated at 400,000,000,000 tons. The giant Stalinsk metal works is the pride of Soviet Siberia. And the "Black Soil" steppe region at the foot of the Altai is among the most fertile on earth and barely inferior to the Ukraine.

We slowly left the steppes behind us. At Krasnoyarsk we crossed the Yenissei, mightiest of all Siberian rivers, with its grim valley which was famous as a convict settlement for two hundred years. At Shuhenskoye, on the upper course of the river, Vladimir Tlyich Lenin lived from 1897 to 1900. Here he married Nadezhda Krupskaya, who was also living in exile. In the little settlement at Kureika the banished Stalin lived, and today its inhabitants point with pride to his old vegetable garden.

The radio has stopped, but two elegant foreign ladies play the latest hits over and over again on a gramophone. A Russian state actress makes a third. The steaming samovar at the rear of each coach is always good for boiling water. Rum and vodka make the rounds. Outside, the silvery moon gleams in the silent Siberian night.

Early in the morning of the sixth day we reached Irkutsk. The

military are much in evidence here, where the domain of the Far
East Army begins. The city is the military, intellectual and economic
centre of eastern Siberia.

At length we reached Lake Baikal, and then came the frontier.
The little Mongolian station-master looked very picturesque with
his copper-coloured face and scarlet cap pulled down over his cute
little slits of eyes. The barber's shop had Mongolian hairdressers.
You could have your whole head shaved for two roubles. Soldiers
and posters swarmed in the station. "Greetings to Soviet youth,
which obeys Stalin's commands!" "Greetings to our young recruits!"
"There is no worse crime than betrayal of your country, undermining your army, spying, helping the enemy or selling military
and state secrets", were among the exhortations. The courts in
the Far East worked swiftly. According to the Mongolian *Pravda*,
there had been two hundred and fifty executions for treason in the
Far East in a few weeks.

Early in the morning our train reached Otpor, the frontier
station. Although very few travellers were going on, the Customs
examination lasted until midday. Then the doors were locked
and the blinds pulled down. Soldiers with fixed bayonets patrolled
the station. It was forbidden, under the severest penalties, to look
out of the window whilst the coaches were being hauled across
no-man's-land into Manchuli. There is nothing interesting to see!
A few watch-towers, some barbed wire and a little underground
shelter here and there. Once through, the breath and smell of
eastern Asia greeted me. A flood of shrieking and gesticulating
Chinese poured into my compartment. My luggage was snatched
from my hands without consulting me. I accepted the inevitable
and myself shouted "Nikitin Hotel", very dubious whether I should
be understood. The mob vanished like a ghost. I met it again, still
up to strength at the door of my bedroom. It was a relief to get rid
of it with a fat yen note.

The passport examination by a Japanese officer was only super-
ficially courteous. With a deep bow and inscrutable smile he wished
me a pleasant journey in faultless English. Manchukuo appears
to be the melting-pot of the Far East. A motley admixture of races
(in Europe there is nothing like it outside the Balkans) means that
stable conditions will never be possible here. White Russians,
Chinese, Mongols, Koreans and Japanese all live together. Communist propaganda finds it fertile soil. We Whites are hated everywhere. And what about the Japanese? Eastern Asia keeps on
smiling. When will this mask fall? Today Asia already belongs to
the Asiatics. Patience is a virtue of the yellow race. *Panta rhei*:
everything flows on.

When, a few months later, I again made an extensive tour in
the Soviet Union and my manifold impressions were no longer a
novelty, Russia, and especially Asiatic Russia with its boundless,
monotonous and elusive distances, seemed to me a sinister, menacing
monster pregnant with danger. Standing on the banks of the
Moskva and contemplating the walled Kremlin—looking grim and
gloomy despite all its Oriental embellishments and shining domes
—I realized that the men in the citadel are thinking in dimensions
and eras different from those familiar to our leaders in Germany
and their even more nervous bourgeois counterparts in the Western
states. While the West was tearing itself to pieces the Soviet was
sitting here in the old fortress of the Tsars and biding its time as
only Asiatics can. There was little sign of any feverish mobilization
or open preparations for war, and yet Russia was armed, better
armed than Berlin wanted to believe. The troops stationed in
Moscow and the Siberian garrisons made a good impression. The
discipline, food and equipment of the Red Army men I met was
very different from what was said about them in the reports appearing in the German and Western press during the 1939–40 winter
campaign in Finland.

The little I saw of the Russian Air Force gave me the impression
that it was antiquated and could not be regarded as a military
factor. Yet, despite all technical developments, Russia had remained a land power of the first order. Her inexhaustible reservoir
of men and her vast distances are her weapons.

As we came closer to the Western frontier of Russia we saw
more and more of the Russian army. The troops seemed to be in
no hurry, as most of them were on foot. But the direction in which
they were marching was indubitably west. The War had diverted
the Soviets from their domestic ideological and colonizing tasks.
Only the process of industrialization was intensified with might
and main. Yet their efforts did not seem to be limited solely to
Russia as their great final goal, world revolution, had come very
near as the result of the "self-destruction of the capitalistic empires".
They considered not Germany only but the whole Western world
as their irreconcilable foe, as was explained to me by young cadets
and students whose unblushing chauvinism, incidentally, was not
in keeping with their doctrinaire Bolshevik education. Red Army
officers who were travelling in the train with me, and whose
knowledge of German and German literature was amazing, lost
no opportunity of showing their glee over the English defeats.

After my return, the more I thought about the limited space
and resources at my country's disposal, the greater seemed the

danger of something much worse than an ideological controversy between swastika and hammer and sickle. But there was little time left for rumination as our Group had suffered heavy losses in the battle with England and could not permanently allow its young officers to range the world as philosophizing globe-trotters. So back to the joystick and leaving politics to those who were paid for it.

CHAPTER XII

WAR WITHOUT PITY

THE 21st June, 1941. *Alea jacta est*. The die is cast. At midday
Adolf Hitler's call to the soldiers on the Eastern Front is being
read. At the moment we are flying against convoys off the western
coast of Scotland. A few minutes before briefing I am handed the
usual blue-green chit with the order to read it out to all members
of the squadron. In flying kit and Mae West, my thoughts concentrated on the sortie ahead, I tear open the innocent-looking
envelope in front of the men. None of us know its contents. We
have not the slightest idea what it is all about. As I slowly read
out the note, it is some time before the full meaning and consequences of this declaration of war upon Russia dawn upon us
and we are still puzzling it out while the lorries which take us to the
Stavanger-Sola airfield are standing with engines running behind
the hutments and telling us it is time to leave.

What had happened to relations between Europe and Asia?
The Russo-German Pact of Non-Aggression concluded on the 23rd
August, 1939, and which was confirmed by Ribbentrop's declaration when he left Moscow ("The Führer and Stalin have decided
for peace") was already in jeopardy in November, 1940, when the
two sides failed to reach agreement in the Berlin negotiations with
Molotov about Russian military and political influence in Bulgaria,
Roumania and Turkey.
The Soviet Union was obviously concerned about Germany's
unexpectedly swift victories. They too adhered to the strategic
ideas of the First World War and had hoped that the warring
nations would grind each other down in long and wearisome
trench warfare on the Westwall (Siegfried Line) and Maginot
Line without final victory falling to either side. That speculation
emerges clearly enough from a secret circular issued in the spring
of 1939 by the Chairman of the Comintern, George Dimitrov. In
it he wrote to all the foreign sections of the Communist Party:
"The Soviet Government and the Party have decided that the
best course is to keep out of the conflict, but be ready to intervene
when the warring powers are weakened by the war, in the hope
that we may bring about a social revolution."
So in the spring of 1939 the Soviet Union was sure of its war,

which Dimitrov described as a means to the end of world revolution. At the same time, Stalin, through Astachov, the Soviet Trade Attaché in Berlin, offered Hitler closer economic co-operation, a step which confirmed the latter in his unyielding attitude to Poland and the powers which had given her a guarantee. Only a simpleton could have believed that these mutual manifestations of Russo–German friendship were genuine.

Hitler made his first preparations for the war with Russia in the early autumn of 1940, about the time when the German Luftwaffe began the Battle of Britain. Its speedy failure had shown that its fighting strength against Russia would be sorely diminished. The two air theatres would impose on German air strategy a division and dispersion of effort. The result was that numerically inadequate forces were available to a strategy seeking a quick decision in the East. Squadrons had to be withdrawn from the West. The intention to return them to the West in quite a short time was and remained a piece of wishful thinking, as foolish as the dream of a blitz campaign.

The Luftwaffe entered the war against Russia with, at most, 2,800 effective bombers, dive bombers, fighters and destroyers.

The Russian Air Force against them was seven times bigger, even if it was inferior in quality to the German. But the Air Staff would not realize its actual size and possible growth. Information on the subject was always rejected or minimized. Its full strength was given as, at most, 5,000 aircraft of all classes. When we remember that Hitler, in his speech in the Reichstag on the 11th December, 1941, spoke of 17,322 destroyed or captured Russian aircraft (and even the true figure was 8,500) it showed how badly the Air Staff deceived itself.

The area in which the Luftwaffe would have to operate seemed a nightmare to the ordinary man, seeing that it meant a 1,000-mile line from the mouth of the Danube to the Baltic. In the first sector, extending to Leningrad and Moscow, it had 600,000 square miles to cope with. North of Leningrad there was another front. It extended from Lake Ladoga 620 miles west to the White Sea and then past the Barents Sea to Murmansk and Norway. The important industrial centres, moreover, lay 125 to 250 miles east and north of Moscow, though the second and stronger heart of Russia's war economy beat on the lower Volga and in the Caucasus, the Urals and, above all, Siberia.

The Russian spaces might have been a giant crater, though their character varied. Had anyone thought of the great variations in temperature, or of the massive dust clouds over the roads and tracks? Did we know of the cloudbursts which in a few moments

made the roads impassable and the rivers overflow, carrying away
the bridges? Mud in spring and autumn made swift movement
impossible and brought war of movement to a stop—that is war
of movement by Europeans, not by Russians. Had all that been
allowed for? No, for we shall have won long before the winter!
Russians were simply regarded as subhuman. The real Russian
enemy were supposed to be concentrated round Moscow. But
though Moscow is the modal point of communications, Russia
is not just Moscow. Asia too marches under the banner of the
hammer and sickle.

It was under these auspices that the war against Russia, the
clash between two ideologies and their dictators, Hitler and Stalin,
began on the 22nd June, 1941. Before it was over Hitler won many
battles but lost it. Luftwaffe and army alike fell victims to the
immensity of its wide-open spaces.

The four first phases of German victories and invasion take us
to the opening of the attack on Sevastopol. They can be analysed
briefly thus:

22nd June to 21st July, 1941.

In the first few days command of the air over the operational
areas was won—and not relinquished. Thanks to unceasing air
attack, the army was supported against the Russian ground troops
in action and its rapid advance was made possible by our onslaught
on the lines of communication, particularly railways. The direct
participation of the Luftwaffe had a great deal to do with the
envelopment at Bialystok-Minsk and was the preliminary to those
at Smolensk, Gomel and Uman.

22nd July to 9th October.

This is the period of the diversions to north and south, while
the central sector was temporarily quiescent. The work of the
Luftwaffe in the north carried the German front to Leningrad
and ensured the conquest of Estonia, notably Reval and Narva.
The process of actively supporting the field army in action was
continued. On the central sector the army girt up its loins for new
operations. Russia was occupied in building up her reserves.
In the Black Sea we flew sorties against ships and harbours. We
also bombed Leningrad and Kronstadt. But these tactical successes
and our weak attacks on industrial centres such as Gorki, Rybinsk,
Tula, Kharkov and Yaroslavl, had no effect on that fulcrum of
the defences which was the Moscow area. The dispersion of our
effort in the Russian theatre began to be noticeable.

9th October to the beginning of December.

The Battle of Moscow. The Luftwaffe was distributed over the
whole front and gave valuable support to the army thrusts which

led to the Battle of Briansk and Vyasma. In the far north the attack on Murmansk made no progress. The Murman railway was often attacked, but without permanent effect. In the central sector we concentrated on the enemy ground troops and their supply lines, but without materially hampering the fresh Russian concentration. The strength of the Luftwaffe was increasingly dispersed over many targets. The 2nd Air Fleet under Field-Marshal Kesselring and the IInd Air Corps were transferred to Italy, further weakening the Moscow front. The movements of the army and the Luftwaffe were bogged down by the arrival of the season of mud and an early and extremely cold winter. German losses were heavy.

6th December to March-April, 1942.

On the 6th December began the Russian counter-attack in the Moscow area. The resultant loss of ground made a decisive German counter-blow against Moscow impossible. The transfer south of the main German effort could be observed, but Sevastopol had not yet fallen. The Luftwaffe's task of assuring supplies from the air had been inaugurated in a big way at Demiansk and Cholm and various smaller operations of the same were also in progress. In spring the attack on Sevastopol was prepared, preliminary to the offensive against Stalingrad and the Caucasus. It was too late for them to succeed.

In the spring of 1942 the Air Chief of Staff sent me to the Crimea and Nikolaiev to join the bomber squadrons there in attacks on ships in the Caucasus harbours and Sevastopol. At that time my group was considered as the most successful in attacks on enemy shipping. Then I was given various special jobs in the western and southern theatres. A few days before the storming of Sevastopol I was summoned to headquarters by Jeschonnek. I was to return to the Crimea.

Extracts from my diary, June-July, 1942.

"Here I am back in the Crimea.

"The heaviest rocket bombs are all ready in case the Russian fleet tries to relieve the fortress of Sevastopol, now isolated.

"My old Ju 88 has brought me non-stop straight across the Balkans from Italy. We came down from 16,000 feet to our destination, Eupatoria airfield. The heat was almost unbearable. It brought a return of the fever I have only just got over, and with it a sort of mental paralysis. I was glad enough of my thin knaki uniform. I took a Fieseler Storch to visit the headquarters of the divisional general commanding the VIIIth Air Corps, whom Jeschonnek had informed of my coming. I came in low into the valley of the Ishuruk-su in which lie the town and Khan's palace of Baktshisarai, 'the house of the gardens', then headquarters of Colonel-General Baron von Richthofen.

"I climbed again until I received a green light signal. There was a Volkswagen waiting. It was late afternoon and the air like a hot bath.

"We drove through the high walls of the Khan's palace in the centre of the town. It had previously been a museum. An A.D.C. met me in the garden. We hurried through a whole series of halls and rooms and a pergola on the sunny side and after a short period of waiting I found myself in the presence of Colonel-General von Richthofen.

"Seated at his desk, his back to the wide-open window, Richthofen almost looked a Khan himself, with his high cheek bones, small, narrow eyes and weather-beaten features. Our talk was brief and military. He explained the attack on Sevastopol, which had just begun, and outlined what was required of the Luftwaffe, which was to smother the fortress with bombs. I was to visit the air strips involved the same day.

"Richthofen seemed to be in his element. It was a job after his own heart. He was one of the most striking figures among the Luftwaffe leaders in the war. His friendship with Jeschonnek, the Chief of Staff, must have made things easier for him.

"Next morning I flew over Sevastopol myself. Bombs at the feet of the army—a *sine qua non* as it had been before. When the army wanted a decision in a battle there were loud shouts for 'that whore, the Luftwaffe'. But when it was a matter of deciding a whole campaign, such as the Moscow offensive in the summer of 1941, the idea was not carried through to its logical conclusion.

"But was the taking of Sevastopol really of strategic importance? To the German leaders it seemed more a question of prestige and a political factor against Turkey. It was certainly to Russia's interest to hold it, though it is probable that they had 'written off' the garrison. At that moment it must have suited them to tie down big German forces there so that they had to postpone the next part of their programme. We may conclude that the Germans were led into a trap when they set out so late to capture the fortress.

"In all we had assembled about 400 Stukas, bombers and fighters on the airfields in the vicinity of Sevastopol. It meant that we had to keep 200 to 250 aircraft always ready for action every day.

"From the air Sevastopol looked like a painter's battle panorama. In the early morning the sky swarmed with aircraft hurrying to unload their bombs on the town. Thousands of bombs—more than 2,400 tons of high explosive and 23,000 incendiaries—were dropped on the town and fortress. A single sortie took no more than twenty minutes. By the time you had gained the necessary altitude you were in the target area.

"With all the smoke and dust, amid the roar of the detonations, the battle area is largely invisible to our troops on the ground, though they could see the bombers fly down into the wasps' nest which is the shrinking defence ring. The screaming descent of the Stukas and the whistling of falling bombs seemed to make even nature hold her breath. The storming troops, exposed to the pitiless heat of the burning sun, paused for the few seconds which must have seemed an eternity to the defenders.

"The Russians clung to their mother earth with unparalleled obstinacy. If no other way lay open, they blew up their forts and defence works, often a long way underground, together with their assailants and themselves. The Russian A.A. was silenced in the first few days so the danger to aircraft was less than in attacks on the Caucasus harbours or Russian airfields. Yet our work at Sevastopol made the highest demands on men and material. Twelve, fourteen and even up to eighteen sorties were made daily by individual crews. A Ju 88 with fuel tanks full made three or four sorties without the crew stretching their legs. It meant tremendous wear and tear for the aircraft and the ground staff, those unknown soldiers who could not sleep a wink in those days and nights and were responsible for the safe condition of their machines.

"Under the massive weight of the bomb carpets, the heavy artillery of the army and the 'Thor' super-mortar, even the most desperate defence was bound to break down. Day by day the ring got smaller and smaller. Thousands of German and Russian soldiers died in fierce hand-to-hand fighting. The earth drank in streams of blood and sweat while in the old palace of the Tsars at Yalta the army chiefs prepared to celebrate victory—incidentally a celebration which was to be rudely disturbed by a Russian air raid.

"The only times when there was a short pause was when the sun sank behind the Black Sea, its last rays bathing the fortress and harbours in a blood-red glow. And only when the last Russian soldier had fallen in the Chersonese or surrendered in the lighthouse did the end come on the 4th July, 1942.

"Such was Sevastopol, a name spelling something gruesome and horrific to all who were there. Attacker and defender alike fought with a fury which was quite exceptional even for this war."

To the Russians this struggle was a demonstration, not only

of their fanatical will to resist but of growing power also. To the Germans it was the high-water mark of their victories in the East. From now on it was downhill for them.

Göring marked the end of 1942 with these words:

"The German nation knows that once again it was the Führer who made these victories possible. The genius of his leadership has been the basis of all our success. In his own person he unites the art of the statesman and the gifts of the great commander. The Führer produced the brilliant plan of campaign by unremitting thought and toil. Germany had one great man of that kind in Frederick the Great. Providence has given us another in Adolf Hitler. The boldness of his schemes and plans is unique. . . . Hail to Adolf Hitler, our Führer and Commander-in-Chief!"

But at that moment was it not true that the skies of Europe had long been heavy with menace? The signal had already been given with the abandonment of the Caucasus, the isolation of an army at Stalingrad and the retreat to the Don.

And yet Hitler had nearly won his war without pity. When the land operations carried the small spearhead of our advance far into the Crimea and our vanguard reached the Caucasus and German soldiers planted the German flag on the very summit of Elbruz, our political leaders firmly believed that the way to India lay open, and preparations were actually made for an advance into the lands of the Thousand and One Nights and the White Elephant.

And then came the terrible calamity. Stalingrad.

Neither the German communiqués nor the explanations of Dr. Goebbels had much regard for the truth. Just as the taking of Sevastopol, the "impregnable fortress", had been celebrated as a feat which decided the campaign in Russia and the retreat from the Caucasus described as a "withdrawal in accordance with orders" Stalingrad was said to be a "slight anxiety". Those who drew up the communiqués were silent about the real figures of dead, wounded and missing. Hardly a family in Germany was spared the consequences, direct or indirect, of that fateful enterprise.

Marshal Timoschenko's first great offensive which cost us Stalingrad was followed by the second, in the Briansk-Kharkov area, in the summer of 1943. Smolensk was the second target of the Russian armour. The Donetz basin was lost. The Russian steam-roller forged ahead.

Now was the moment to draw the logical conclusion. There was still time for a great strategical air attack on the Russian power stations and their vulnerable turbine installations, which had partly been constructed in Germany and would have been very difficult to replace from Russia. Such an onslaught would very probably have had a decisive effect on the Russian power output. The main power stations were the sole vulnerable point in Russia's defensive system and their elimination meant that her armament industries would be paralysed. The High Command had no economic maps on their tables while Hitler was pinning little flags representing individual companies and batteries on the situation map of the Eastern frontier.

The generals, with Field-Marshal Keitel at their head, thumbed the seams of their trousers like a lot of recruits and stood in silence when summoned to receive orders. When someone reared up— like Colonel-General von Kluge, proposing a withdrawal in view of his shrunken, composite "divisionlets", when he was hard pressed by the Russian counter-offensive at Rjev—Hitler was not to be moved from his decision: "All right, Kluge, you'll have to hold the position with your divisionlets!" Further talk seemed useless. Even a quick-witted Berliner like Kluge held his tongue.

Was it surprising that many of the generals and officers of the General Staff ceased to believe in victory?

Colonel-General Jeschonnek died by his own hand on the 19th August, 1943. Göring had promised that Stalingrad should be supplied by air and when he could not keep his promise, retired to Karinhall and left the responsibility to Jeschonnek. Although one supply sortie had followed another it proved impossible for the Luftwaffe effectively to supply an army which was surrounded. At about the same time the heavy Allied air raids on German cities and war factories began. Jeschonnek, who had always striven to meet the wishes of the army in the East, was now made the scapegoat for the Luftwaffe's failures in home defence also. Bitter words passed between him and Göring.

From my diary:

"On 18th August, 1943, just before 11 p.m. I am rung up in my room in hospital in Berlin. First priority call: Jeschonnek calling. 'Well, how are you? Have you forgotten all about me? I'll have to send a hospital plane to fetch you here.' 'I'll be up and about in two days, Herr Generaloberst, and will come over at once.' 'No, no, take your time. You must get well first. I only wanted to know how you were getting on. The best of luck. Keep fit.'

"That was all. At the other end the receiver was replaced. My friend, Major von Harnier, who was visiting me, had listened in. We looked at each other in dismay. I had never heard the General-oberst talk like that before. Next morning I rang up Leuchtenberg, Jeschonnek's A.D.C. 'You needn't come now,' he said, 'the General-oberst is dead; tell you about it later.' "

On the 20th August, 1943, the front page of the *Volkische Beobachter* carried the following article:

Death of General Jeschonnek:
The Chief of Staff (Air) dies after a serious illness.

On the morning of the 19th August Generaloberst Hans Jeschonnek, Luftwaffe Chief of Staff, died at Luftwaffe headquarters, after a serious illness. His death means a particularly cruel and painful loss to all the services. The Luftwaffe loses an outstanding soldierly personality who prepared the way for great military victories in many campaigns.

"As the closest and most loyal colleague of the Reichsmarschall, Colonel-General Jeschonnek, occupying a post of the highest responsibility, gave his life for Führer and Fatherland. The very embodiment of devotion to duty, and without consideration for himself, he knew nothing except the great goal of victory. Even an insidious malignant disease was totally unable to paralyse his inexhaustible energy up to the very last moment. We lower the colours to the General who has been summoned to join the Great Army. The name of our tried and trusted Luftwaffe Chief of Staff will never be forgotten."

Comment would be superfluous.

In his last will Jeschonnek had declared that he did not want a public funeral. He wanted to be buried like a private soldier at "Robinson", his headquarters at Lake Insterburg in East Prussia, where he had been working for more than two years. How often had he and I strolled, or sat on the wooden bench, in the little garden by the lakeside while he poured out his heart, a heavy, weary heart, to me, a man scarcely twenty-five years of age, and almost frightened at his confidences. I had become the thorn in the side of the General Staff. Jeschonnek had contemptuously tossed these intrigues aside. He was the Luftwaffe beast of burden and Göring's whipping boy when anything went wrong. But he took all that in his stride, as a soldier of the old school. Then things went from bad to worse with the Luftwaffe and the long-standing differences between him and Göring, which had been fanned by two persons in the latter's immediate entourage, came to an open breach.

Jeschonnek saw no other way out than a bullet. While his staff were out at the morning conference he shot himself in his little blockhouse. The background to his suicide is one of the darkest chapters in the history of the German Luftwaffe and its leaders.

Jeschonnek is a perfect example of how honest soldiers with the strongest possible sense of duty were convinced by Hitler. His sober views and biting irony were well known. He never concealed from me his disgust at Göring's way of life. His contempt for Milch was

proverbial. He was utterly modest and self-effacing personally—
yet Hitler's orders he obeyed without question. Could he escape
his responsibility by a voluntary death? Would his sacrifice avail?
Could it shake up our leaders and open their eyes to the menacing
internal dangers?

It was too late.

In August the newly formed Russian armies, once more strongly
supported by modern tanks, thrust forward to Odessa. American
munitions deliveries via Murmansk, Archangel and Persia, made
their presence felt. The German 17th Army was destroyed. Führer
Headquarters repeated its order that the German troops must not
abandon the Crimea. At Tscherkessy a large German force was
surrounded and the Russian spearheads approached Gomel. The
Pripet marshes, well known from the First World War, were
reached and the German front was thereby split in two. The Russian
masses seemed to be inexhaustible. With six army groups and six
armoured armies the enemy opened their gigantic winter offensive
of 1943–44 and simultaneously relieved Leningrad. In January,
1944, the Polish frontier was reached. Russia was firmly in Russian
hands again.

"Where is the Luftwaffe?" had become the despairing cry of
the German army, and it was understandable enough. But the
Luftwaffe had already been worn away in the fighting on land.

The German soldier had never been trained for a defensive war
and was not even allowed to fight one. The orders came from
Führer H.Q. The army commanders were deprived of the last
shreds of independence. Hitler had long ceased to trust the army.
Raising the new S.S. formations always had priority over army
needs. The weapons and equipment of the S.S. were perfect. It
took the pick of the depots. But its commanders were not infrequently
without much military experience and it often happened that this
defect was not made good by higher morale and fighting qualities.
So the losses of the Waffen S.S. were particularly high.

Hitler's visits to the men in the front line became increasingly
rare and at length ceased altogether. The 20th July, 1944, saw the
attempt of Colonel Stauffenberg to assassinate him at the "Wolf's
Bunker", the Führer's Headquarters in East Prussia. Hitler escaped
with his life but received greater injury than was made public at
the time. One thing should be known if what happened afterwards
is rightly to be understood. The new Luftwaffe Chief of Staff,
General Korten, was killed.

A few weeks later Hitler said at a situation conference that
he had received a memorandum from a well-known bomber officer
and it had opened his eyes to many things he had known nothing

about before. Two days later I was summoned to the presence of Reichsführer S.S. Himmler and the same evening went to Führer H.Q. "On the Function of the Luftwaffe in this War", was the heading of my memorandum. It did not mince words. When I passed through the second barrier of the Führer Bunker two hours later I knew that it was too late for a change. At this conference, as at all others I had with him, Hitler listened quietly and let me say my say. He seemed to understand my arguments. Six months later he reminded me of my proposals. At that time I was at the Chancellery almost every day. Too late.

Meanwhile there had been a fresh disaster on the Eastern Front. More than 160 infantry divisions and four armoured armies, with powerful support on the flanks, had destroyed the German Centre Army Group of some 20 divisions. At the end of July the Russian infantry and tanks were at the frontier of East Prussia. Finland made a separate peace. In August the last German counter-offensive north of Warsaw had no particular success. The revolt in Warsaw was smothered in blood and Soviet Russia refused all help to the Poles. On the 2nd October, 1944, Warsaw capitulated.

The Russian steam-roller rolled on to the West. On neither side did the air force play a decisive part, though the Russian offensive had good support from the air. The German armies were increasingly split apart. The Italian front could not be held and the Anglo-American invasion of France succeeded. In this situation Hitler decided on his Ardennes offensive.

Up to January, 1945, Germany, now surrounded, was granted a last breathing-space. Then the Russians were on the Oder and Neisse and the Anglo-Americans crossed the Rhine. The last resources of the German army, air force and navy were scraped together to hold up the Russian advance. Dive-bomber and medium-bomber crews under the command of Colonel Hans Ulrich Rudel, the bearer of the highest German decorations, and what else was left of the Luftwaffe, offered a last, desperate resistance at the gates of Berlin.

On the 5th March, 1945, the following Führer Order was handed to me:

To the C.-in-C. Luftwaffe, High Command Luftwaffe, Führer H.Q. Staff, Chief of Staff Army.

Transmitted to Lieutenant-Colonel Baumbach, Commodore of No. 200 Bomber Group, on the 5th March, 1945.

1. I order Lieutenant-Colonel Baumbach, Com^e 200, to bomb all the enemy river crossings, including Oder and Neisse.

2. Lieutenant-Colonel Baumbach is to employ all suitable resources, whether Wehrmacht or industrial, and make the necessary arrangements.

9

3. He is responsible directly to the C.-in-C. Luftwaffe and will be employed in the command area of the 6th Air Fleet.

4. His operation orders will be issued by the C.-in-C. Luftwaffe as agreed with the C.-in-C. Wehrmacht.

Signed: Adolf Hitler

Once again I was in the shattered conference room of the Chancellery and accompanying Albert Speer to the Führer's little air raid shelter. We said, orally and in writing, that the war was lost and that it was the task of the country's leaders to save the German people from mock-heroic annihilation. It was in vain. Hitler relied on providence and Wenck's spectral army. But he remained in Berlin. At the end of April the capital was surrounded by the Russian army. On the 2nd May the Red Flag with the Soviet star was flown from the Brandenburger Tor.

The Eastern Campaign, the war without pity, was over.

CHAPTER XIII

THE FIGHT FOR THE MEDITERRANEAN

WHILE the campaign in the East was approaching its decisive turning-point the Anglo-Americans appeared on the scene with their invasion in November, 1942. In the Mediterranean area the Axis powers had nothing to match their material superiority. Anything that could be spared from other fronts was sent post-haste to the battle for the Mediterranean.

Extract from my diary:

"For two days our aircraft hung between heaven and earth on their 3,100-mile journey from the North Cape to the Mediterranean.

"When we came down in a wide sweep to Comiso airfield, the snow-capped Etna lay before us between the burning sky and the quivering dust-choked tarmac. The sound of all the aircraft of our squadron flying in was echoed from the nearby hills. To the south the sun-bathed horizon glimmered between gnarled olive trees and withered cactus hedges, and there was *Mare Nostrum*, a steamy haze, a great question mark. When we switched off, Sicily's midday heat enveloped us like a furnace and reduced us to silence.

"It was the 10th November, 1942, and for forty-eight hours great convoys, carrying troops, equipment, munitions and oil and protected by British warships, had been helping to extend the bridge-head at Algiers, now the most important port of western North Africa.

"Sitting on my pack in the meagre shade, I held the tattered pages of my diary between my knees, while the golden rays of the sun overhead eliminated all life and colour just as the moon does. The sweat poured down us and our clothes clung like glue.

"At the edge of the airfield we could see a three-storey white building which half concealed a squat repair hangar and two montage hangars, surrounded by a large number of fighters and bombers apparently in need of repair. The narrow, battered ring-road seemed to get lost at the far end of the tarmac, which was once more a leprous, dead expanse. Around me was a modern encampment of antediluvian warriors, swathed in thick angora shirts, gay shawls, fur-lined tunics and fur caps. Captain Dahl-mann's Airedale terrier, Pax, frolicked round the baggage strewn about and left his usual trademark everywhere. We waited.

"We waited, and no one paid the slightest attention to us on that first day on Sicilian soil. My mind gradually became as empty as my dreams."

What had been the course of events between Italy's entry into the war at the end of the "Blitz" campaign in the West and the great Anglo-American offensive in November, 1942?

Fascist Italy's declaration of war made the Mediterranean an operational area. As in the battle for England and the Battle of the Atlantic the Axis powers came up against British sea power. England must keep open the Gibraltar–Malta–Suez route to India. Italy's own forces were quite inadequate to cope with British sea and air power, with its strongly-fortified bases, by themselves. It is true that the Italian navy comprised modern warships, including submarines, and in the *Regia Aeronautica* the personnel of the fighter and torpedo aircraft was good. But the aircraft themselves were out of date. And the outstanding question was: would these two arms of the service really work together?

The Apennine Peninsula is extremely exposed to naval and air attack. Sicily and the island of Pantelleria are outflanked by Malta. Tripoli is a blind alley. It was a mistake to leave Bizerta in French hands at the time of the armistice with France. The European southern flank could be secured only by closing the Gibraltar–Suez route. The military problem to be solved was— Mediterranean or the Suez Canal? Instead, Rommel marched into the North African desert.

Crete should have been the air base against Alexandria. An offensive focussed on Gibraltar and the capture of Malta would have been the second task in the Mediterranean. It would have been a very serious danger to the mainland of the Near East, Arabia, Irak, Palestine and Transjordan, which were under British influence. The capture of Malta would have meant the loss of the most important base on the Mediterranean route. Gibraltar could have been made a mousetrap.

At quite an early stage England's position in the Mediterranean could have become most precarious if there had been co-operation in those strategic designs between the German and Italian air forces—with expert leadership, supported by parachute, glider and naval formations. But the complications of dual command in the Axis were always giving the British sufficient breathing space until American support arrived and sheer weight of material decided the issue in the Mediterranean. What Russian weight of numbers accomplished in the East was matched in the Mediterranean by weight of material.

From a strategic point of view Rommel's North African campaign was an operation simply hanging in the air, as neither his line of retreat nor his supply line was secure. The Afrika Korps could not be kept going by a few ships and inadequate transport by air. Yet "the tanks are forging ahead in Africa"!

Rommel sped eastwards at the head of his tanks. The names of Derna, Tobruk and El Alamein were always cropping up. Often enough his staff did not know for days where he was or how far their tank spearhead had got. Within a very short time, the name of Rommel, as the mighty thruster and master of improvisation, had a magic effect on friend and foe alike. "There's another Rommel on," people used to say when another victory in the desert was announced.

Field-Marshal Erwin Rommel was not only the outstanding tactician—the man of expedients—but also the most chivalrous soldier of the North African campaign.

The concentration was supported by a dive-bomber Group. For a short time Marseilles was the brightest star in the fighter-pilot's sky. Bomber squadrons pasted Malta. When the last enemy fighter had disappeared and all was set for the invasion of the island it was called off. Only individual targets such as merchant vessels, convoys and warships were attacked. At night there were sorties to Alexandria. These were all pinpricks without strategic importance. Then we resumed protecting our own convoys with aircraft and the miserable remains of our air strength was consumed in Rommel's offensive. Just as the Luftwaffe was pulverized in the ground fighting in Russia, the flower of its young officer corps was sacrificed in the Mediterranean.

The C.-in-C. South and Commander of the IInd Air Fleet, Field-Marshal Kesselring, was brought prematurely from the Russian front to the Mediterranean. He had not been an airman originally, though personally he was one of the most upright and attractive figures in the German services, the last knight of the Luftwaffe, in the best sense of the word. In more than a hundred sorties, utterly careless of his own life, he had kept his squadrons together in the most desperate circumstances. But even he was a prisoner of his training and experiences in land warfare. As commander of the Luftwaffe, he could not deal effectively with Hitler. He could direct strategic air warfare. He knew all about it. Into the scale he threw all those high personal qualities which made him acknowledged by friend and foe as the most striking figure of the war in Italy and the Mediterranean. But he did not prevent the Luftwaffe from bleeding to death in the Mediterranean. Hitler and Rommel were too much for Kesselring.

And what had happened so often before was repeated. Rommel nearly got to Alexandria—though his army was out of breath and had no reserves of any kind. A near thing. At that time Rommel was the only source of victories for the communiqués and German propaganda.

Meanwhile the Americans had set up their air transport route to Egypt via Natal, Monrovia, Freetown and Fort Lamy, and made their preparations for an invasion of North-west Africa, while the British opened their offensive from their Egyptian base against Rommel's army.

The threat of this two-front Anglo-American attack had for some time been hanging over the North African sky like a sword of Damocles poised to destroy Rommel's troops, which were exhausted after being driven to the limits of human endurance.

To anyone involved in this murderous struggle it was obvious in a few days that the geographical situation and the disparity between numbers and resources on the two sides spelled the beginning of the defeat of the Axis powers in the Mediterranean. America was to prove not only an industrial colossus but an inexhaustible industrial colossus. The German leaders again refused to see the approach of a catastrophe which was practically contemporaneous with Stalingrad and transferred the initiative to the Allies.

The day before the 9th November, the anniversary of the 1923 affair, Hitler met his old party comrades in the Löwenbräukeller in Munich and minimized the turn of events at the front in the following fashion:

"Roosevelt may be going through with his attack on North Africa, but we need not waste words on this old gangster's lies. He is without doubt the biggest hypocrite in the mob against us."

It had not been possible for the German submarines and Luftwaffe to spot the approach of the Allied invasion force in time, much less prevent the landings. What we had available was much too small for any effective action, in the absence of reliable preliminary reconnaissance, to have much chance of success. Yet the navy chiefs described the situation in those days as follows: "In view of our inadequate reconnaissance facilities countermeasures could not be taken in the Gibraltar area, but the enterprise will soon be frustrated."

The successes of our submarines off Capetown and Canada were being described as a "promising development", just when the preparations for the Anglo-American invasion of North Africa and its subsequent landings were bringing about a fundamental change in the strategic situation, even on the naval side.

What was the tactical position?

At 1 a.m. on the morning of the 8th November, 1942, the landings in the French North African harbours began, the largest scale landings up to date. More than 850 landing craft put 140,000 troops with the most modern equipment on shore.

On the 7th November, forty-seven American transport aircraft had flown 1,400 miles along the coast to drop parachutists to help the invasion fleet from the land side. The German wireless had on the same day reported enemy aircraft flying from England in the direction of Africa. Within seventy-six hours of the first landings the Allied troops firmly controlled 1,300 miles of African coast. In places the French offered a heroic resistance, but Allied bomber attacks put an end to any serious opposition. Some coastal batteries hoisted the white flag after firing a few rounds. The surprise was complete. Morocco and its main centres, Agadir, Mogador, Safi and Casablanca, and Algeria on both sides of Oran and Algiers were quickly captured. The Eastern end of the bridge-head was thrust forward to Philippeville. With the precision of a watch proceeded the preparation of the Allied forces for the air and ground operations which had been planned. It was an unstoppable avalanche of men and material, in comparison with which the desperate efforts of the Axis powers seemed puny.

In the early hours of the 11th November, German and Italian troops entered the hitherto unoccupied zone in France "to counter the continuation of the Anglo-American attacks effectively", as Hitler wrote to Pétain. The French fleet in the harbour of Toulon scuttled itself. In Marseilles German flak formations took over protection of the merchant ships, now placed under German orders. For the German Luftwaffe a few useful airfields in Southern France provided a new base for the air war in the Mediterranean. If we had only had the aircraft!

Catania, Comiso, Trapani and Gerbini were the airfields in Sicily. Sardinia and Southern France were the depot areas. In the months of November and December the effective battle strength of the German bomber squadrons was 100 to 200 aircraft. Within a few weeks the number was reduced to barely fifty operational aircraft and crews. The small reinforcements that arrived could do little to change the picture. Before long we had only the staffs and skeleton organizations on the airfields—without aircraft, crews or hope.

In the Operations Room of the IInd Air Corps and the 2nd Air Fleet the staff went on operating with groups and wings which existed only on situation maps. In reality, the groups had less than ten crews.

While this was going on, in the eastern Mediterranean some twenty aircraft of the Xth Air Corps were operating, mainly from Crete. These aircraft often landed with literally the last drop in their tanks, as the sorties reached extreme range. They were given additional targets every day.

Rommel's army was crying out for petrol. The few transport aircraft which the Axis powers possessed required escort. The seas had to be searched for the periscopes of enemy submarines. There was a steady call for the bombing of the enemy's fighter and bomber bases which were getting further and further east. The roads and supply routes of inland Algeria had to be strafed with bomb and machine-gun. Malta, the naval and air base, was repeatedly attacked by night. Warships and merchant ships bringing vast quantities of war material had to be sunk. Unloading in port must be prevented. In all these ways the German leaders squandered their last resources.

The operative units of the Luftwaffe soon had a foretaste of the collapse of the Mediterranean front. A further extract from my diary:

"A week later.

" '*Lasciate ogni speranza, voi che entrate!*' The angel of death cast its shadow over us immediately on our arrival. Dante's Hell is a reality here. Roth's crew gone, Lieutenant Grigoleit shot down, little Quisdorf missing, Stoffregen down and mortally wounded, Metzenthin's and Lieutenant Harmel's Ju 88s shot down in flames over Algiers.

"Yesterday evening I had the jumbled-up crews lined up before me once again. Briefing; target same as yesterday. With everything we have. Same as yesterday. Always same as yesterday. And early next morning, when we assemble after the sortie, two, three or four crews will be missing once more. Have not returned from the sortie! Who will be the next? What cruel fate has already decided on the next batch? All of us have got together our few belongings and tied them up as 'casualty pack' by the time the lorry arrives to take us off to the airfield.

"I often lie on my bed in a sort of paralysis, dripping sweat and yet feeling frozen to the bone while gazing at blood-red oranges hanging in the leafy trees. It is worry, heart-rending worry, which shakes me to the core and is growing to a horrible, nerve-racking fever. Since we have been here I have stopped talking to the men. I could not find anything to say which would lessen the feeling of hopelessness. They understand me. No one is quicker in the uptake than the man in the ranks."

Two weeks later:

"I have had a long struggle to make up my mind. It is our last chance. I shall write to Jeschonnek. He shall at any rate know what is going on here. I sat at my typewriter until after midnight, getting out my report. The courier leaves for H.Q. in East Prussia in the early hours.

"I believe I have recovered my old self-confidence. I had nearly lost it altogether in the last few weeks but I shall need it more than ever now.

"I went for a long swim this morning. On the way back I stopped at the cottage of a very old blacksmith I have often called on. Early though it was, he was hard at work when I walked into his tiny smithy in the half-light. His words poured over me like a melodious waterfall. He proudly showed me the sort of things he made—candlesticks and church candelabra of fine workmanship. I felt that it was the last time I should see him and bought a little *objet d'art*. When I was having breakfast with my crew the Ju 88 carrying my report to the Chief of Staff droned over our heads on its journey north."

Major Baumbach.   Comiso (Sicily).
                                        12th December, 1942.

Secret.

To the Chief of Staff, Luftwaffe
    Colonel-General Jeschonnek
        Robinson-East[1]

I have the honour to send the report and appreciation enclosed and have taken the liberty of circumventing official channels. I consider it my duty to give my views on the situation as I see it and I accept full responsibility if I express myself in harsh terms.

The statements I make are based on the activities of the bomber wing which I have commanded between the 10th November and the 10th December, 1942, as well as the experience of other units operating in the same area. I have also had regard to the conditions we found on arrival.

I intended to send in my report earlier. It seems to me imperative now that General Lorzer, commanding the IInd Air Corps, has shown that he has no understanding of our situation.

1. The work of the bomber wing in the Mediterranean.

The wing, comprising thirty-nine Ju 88s, came from north Finland and arrived at Comiso on the 10th November.

Although the IInd Air Corps had had at least two days' notice of our coming, no preparations (quarters, headquarters, clothing, feeding, telephonic network, bays, motor transport, repair and

[1] Code word for Chief of Staff headquarters.

service staff) had been made to receive us. Our arrival was a complete surprise to the ground staff. The ground staff commandant, Major Rose, knew nothing about it before I walked into his room. He is the only bright spot here (he is a clergyman in civil life). The same thing happened in the case of the other bomber wings. The commanders of the individual units had to do everything, and it was only by willing co-operation among them that we got ourselves into rough and ready order for action. Up to today our own technical personnel (Airfield Maintenance Company) have not arrived. The transport of the most essential key men has proved extremely difficult and would have been impossible without my personal intervention and the efforts of the commanders. In that connection I broke at least ten regulations.

It proved impossible to send a few Ju 52/13ms to Comiso with supplies although a bomber wing flew more than 3,100 miles to a completely new theatre of war inside two days, got 70% of its aircraft operational in the absence of the most essential ground staff, and carried out sorties of 1,800 miles.

Neither on arrival nor afterwards was there much sign of preparatory staff work. So far we have seen nothing of any representative of the Air Corps, much less the General. These gentlemen sit far from the firing line in their offices in Catania or, more recently, the comfortable surroundings of the luxury hotel, San Domenico, at Taormina, and in reply to my official reports on the appalling conditions here have informed me by telephone that I may have acquired my experience in the Arctic but conditions here are different and I know nothing about them and must leave everything to them.

I have the impression, shared with the other commanders, that there is a lack of overall direction and control. It is certainly true that vague and confused ideas prevail about the composition and strength of the formations, and the procedure for keeping in touch with them, very necessary for operations in the Mediterranean, is inadequate. With the best of wills I can see no sign of expert and responsible leadership. I can give details whenever called on. In this area the lifetime of a crew is approximately ten sorties. I have given the sorties and losses in an appendix hereto.

I will refer only to one particular matter—the attack on Malta.

In my opinion, daylight attacks on Malta in present circumstances cannot be carried out without extremely heavy losses, and it is even doubtful whether the aircraft actually reach the target. I am told that at any rate the "attempt" must be made, and that we must "improvise" more. It is with such dangerous catchwords that we are directed from the safe distance of the office stool.

2. Command arrangements.

In this special phase of the operations in the North Africa area, if the boldness of the decisions and the perpetual sudden changes of orders had been treated as binding on the officers actually in charge of the operations, no doubt victory could already be regarded as a certainty!

But the truth is that the wing commander, as the last real brake, often has to act on his own responsibility and disregard his orders. When this did not happen, or could not happen (as in the case of Sardinia), unnecessarily high losses have always been the result.

In daylight strafing (it should be remembered that we have only one machine-gun able to fire forward) of airfields and ships in harbour our crews have given of their best—even carrying out what have been virtually suicide orders—though without fighter escort and in the teeth of violent fighter opposition.

In recent times it has not been unusual for an order to be changed nine or ten times in the course of a single day or night, involving confusion in alerts, loading, fuelling up, and so forth— and all that in connection with some sortie from which no real effect could be expected.

The strain on the men has become intolerable as the result of this state of affairs at the top. Between the 19th November and the 12th December it has so far proved impossible to get a single day's rest; on the contrary, it has been nothing but a succession of alerts.

I have failed to notice any appreciation of the fact that since the entry of the Americans into the European theatre (I include North Africa in it) we are now the poor relations, and only by the most careful husbanding can we preserve what is left of our substance.

It is possible that from the level of a wing commander I cannot see things as a whole. But at any rate I am in a position to consider (to take one example) that an attack involving a flight of 680 miles on a pitch-dark night, without weather forecast or any tactical protection, on ships or a little airstrip near Algiers is disproportionate to any possible success, particularly when there will be moonlight within twenty-four hours!

Of course we must keep in the air. The men must have no illusions on that score.

I must make it clear that in spite of everything the flying and technical personnel have surpassed themselves. A proper acknowledgement to these men would be fully justified.

Herr Generaloberst! After the "War in the North" I find myself unfortunately again compelled to address myself to you in this fashion. It may sound pessimistic, but I am telling you no more than the truth.

In all theatres of war there must be men who are prepared to speak their minds to their Chief of Staff. I have always done so and cannot change.

Herr Generaloberst, help us in our great need!

Signed: Baumbach.

The very next day I was talking to Jeschonnek and then Field-Marshal Kesselring. Immediate help was promised. When I got back to Comiso a red telegram lay on my desk. On the orders of the Reichsmarschall I was to be transferred to Berlin at once. I was subsequently told the story behind all this by Jeschonnek.

On the last day of the year I took leave of my men:

"I am talking to you today for the last time and will keep it short as usual. This hour shall be devoted to your future, not to the pangs of separation.

"I cannot and will not tell you what it costs me to be leaving you. Even sorrow is really a present from heaven. We want to preserve a strong and resolute will and, however hard it is made for us, believe in what is good in men, despite all intrigue and slander. In these hard years we have shared together have we not seen plenty of what is good and beautiful? And has it not been a secret treasure which has given us the strength to face all the vexations and worries we have had to cope with? Fortune and honour in this world are no more than fleeting clouds and running water. It is not only the moon which wanes when it is full. Morality alone endures.

"Just stay as you are, for if you do I need never fear that anyone will ever be able to besmirch your manhood. The essential thing is that a man should stand by his word, whatever it may cost him. We have been doing that for years. How often have we seen in times of utmost stress that the coward, lying finger never dared point at us?

"I should never have been able to resist these forces of corruption if I had not always had your confidence, and felt that every man's eyes were on me. The most recent events have brought that home to me. I cannot say what is going to happen now. I will never violate my conscience. What is right must remain right. I still believe it.

"My successor is the best you can have. Live with him as you and I have always lived and endured together. In my thoughts I shall always be with you. And when you are puzzled about service or private affairs I shall be available and will come to you if at all possible. May a good star watch over your future. To me the years I have spent with you will be unforgettable and strengthen me to face the coming storm. For that I must thank you. For the coming year I wish you a heartfelt '*hals-und-bein-bruch*'.[1]"

Göring had taken the lines of least resistance and promoted me to get rid of me. The G.O.C. of the Corps was also removed. For his "outstanding services in the Mediterranean" he was promoted Colonel-General and appointed "Chief of Personnel Equipment and National Socialist Direction". He was one of Göring's cronies. In Berlin I was put in charge of the tests and development of the new guided bomb and told it was an urgent job.

A few months later I was back in Italy on various missions. There the year 1943 had brought no change. The gaps in the formations were only made good by young, insufficiently-trained crews. A few sorties—sometimes the very first—saw the end of them. The back of our bomber and fighter crews in the Mediterranean had been broken and they never recovered. But on the side of the Allies air power increased from week to week. Flak and day and night fighters controlled North Africa and the convoy routes. Time was working for our enemies and they deliberately allowed for it in their strategic plan to roll up the southern front of the Axis.

At first Eisenhower gathered together all his resources before raising the curtain for the Sicily–Italy drama. And then it was to be two years before the Italian front collapsed. Why?

In May and June of 1943 the preparations for the second amphibious landing were under way. German long-range reconnaissance brought reliable reports of the accumulation of men and material in all the North African harbours. So the landings in Sicily were no surprise. Anglo-American air attacks from their Benghazi–Tripoli–Bizerta bases set about undermining the morale of the Italians. In June Pantelleria fell without a fight.

On the morning of the 10th July troops of the American 5th Army under General Eisenhower and parts of the British 8th Army under General Alexander, twelve divisions in all, landed in Sicily. They were faced by equal numbers of German-Italian defence forces, including the "Hermann Göring" armoured division. Yet on the very first day the enemy secured command of the air and sea above and around Sicily. The Italian fleet lay inactive at Taranto.

[1] Best of luck. (Tr.)

Protected by the heavy guns of the warships based on Alexandria
and Gibraltar, the Anglo-Americans were able to extend their
bridge-heads south of Syracuse, at Cape Passaro and in Gela Bay.
By day and night more than 800 enemy fighters formed a protective
screen over the transports. On the 12th July, Augusta, the only
Italian stronghold in Sicily worthy of the name, surrendered
without firing a shot. In the mountainous north of the island what
was left of the German forces put up a desperate resistance for some
weeks. The withdrawal across the Straits of Messina was a great
feat of arms, especially by the German flak, which put up such a
dense barrage that enemy air attacks were beaten off.

On the 25th July the imprisonment of Mussolini marked the
collapse of the Fascist régime. The weak German forces in Italy
were in a desperate position. Führer Headquarters knew about the
negotiations in Lisbon between the emissaries of Badoglio and
Eisenhower. The tense situation in the East made it impossible to
send Kesselring the necessary reinforcements. It was only in August
that German divisions rolled southwards over the Brenner.

In these critical weeks I was attached—on Göring's orders—
to the staff of the Commander-in-Chief South, Field-Marshal
Kesselring. The new commander of the 2nd Air Fleet, Field-
Marshal Freiherr von Richthofen, joined us. From upper Italy
Colonel Pelz directed concentrated night attacks with "pathfinders"
against shipping off the Sicilian coast and landings thereon. They
were and remained no more than pin-pricks. Our own losses were
still on the same high scale. Pelz was soon recalled to direct the
"retribution" attacks against the British homeland. Kesselring was
now in charge of the ground operations only and was to prove
himself an outstanding strategist in the fighting in the Italian
"boot". Von Richthofen can be said to have kept him company,
as for the rest of his service life in the South he was a commander
without an Air Fleet to command.

What was Eisenhower, the Allied Supreme Commander, doing?
His armies gingerly followed the retreating German forces up
Calabria. In Lisbon the negotiations with the Italian representatives
dragged on for weeks. Eisenhower insisted on the Casablanca
formula of *unconditional surrender* even when Badoglio officially
capitulated after the landing at Salerno. Weeks earlier, General
Cabroni, Chief of Italian Intelligence, had assembled four crack
divisions in the vicinity of Rome and was ready to occupy the capital
with the help of a British parachute division which had been
promised. The Italian General Staff was in close touch with those of
Roumania and Hungary with a view to that common action which
was becoming ever more urgent. These British parachute divisions

were expected not only in Rome but in Bucharest and Budapest. But they did not come. Kesselring was able to send his divisions from the vicinity of Grosseto to Rome and receive Mussolini, who had meanwhile been freed by a daring *coup* by German parachutists.

What did Eisenhower do now? He simply advanced northwards from Salerno with the speed of a steam-roller. Wherever the German rearguards turned to show fight he stopped until his air force had laid bomb carpets to open the next few miles of road to his troops. The Benedictine abbey of Monte Cassino, a jewel of Italian baroque, countless other art treasures, towns and last but not least human lives, were the price of the incomprehensible, rigid strategy of Eisenhower, which was none the less described after the war as "genius". Americans who were there still ask: "Was Anzio necessary?" At that time, we who knew our own weakness and our desperate position in Italy were completely nonplussed at the inch-by-inch crawl up the Italian peninsula. Instead of concentrating all his forces on Sicily, why did not Eisenhower simultaneously seize Corsica and Sardinia so that he could carry out lightning landings from the flank at Genoa, Ostia, Naples and in Calabria? With his absolute command of air and sea, his forces were adequate. Nor did the Anglo-American strategy in Italy delay the decision in the Balkans. In fact it facilitated the Russian infiltration a year later and lengthened the war in Europe. Eisenhower could have finished the Mediterranean war at the end of 1943.

Or did the politicians behind the soldiers have the last word? Later on, Eisenhower, speaking about the Nüremberg proceedings against certain German generals, said that soldiers were always having to clear up the mess made by politicians. Presumably he was referring to the influence of Roosevelt and his advisers, which undoubtedly contributed to the prolongation of the war.

The obstinate struggle for every foot of Italian ground from Reggio di Calabria to the Po went on for twenty-one months and the Allies owed the conclusion of that campaign, which only just preceded the capitulation of Germany, not to their superior strategy but their overwhelming weight of material. The war in the Mediterranean will always remain a glorious page in the history of the German services, just as the generalship of Field-Marshal Kesselring and his officers was completely in the true tradition of the German army.

CHAPTER XIV

FORTRESS EUROPE

AFTER the fall of Stalingrad and the surrender of Tunis in April, 1943, the Axis powers and their satellites were faced with the fight for Europe, opening with concentric retreats to their own home countries. For the Reich and its allies the defensive phase had begun. On all fronts the initiative was with the Allies and that fact imposed on the German leaders a fundamental reconsideration of their tactical ideas.

In those critical months German propaganda coined the expression "Fortress Europe" which in the course of the war was successively narrowed down to "Fortress Germany" and eventually "Fortress Berlin". It was to be shown that this was not a mere vague idea or empty propaganda but a concrete, if strategically exaggerated, notion which inspired German military thinking.

In no phase of the war was displayed so vividly the fateful and tragic adherence by the German leaders to their ideas and experiences of the "battles of material" of the First World War. Verdun was the great model of the "untakable" fortress, and Marshal Pétain's memorable battle-cry *Ils ne passeront pas* continued to inspire the builders of Maginot, Schober and Metaxas Lines and West and Atlantic Walls. To the leading soldiers and politicians current military literature had suggested that fortifications, concrete shelters and defence works meant security, and this belief was not to be shaken despite the experiences to the contrary gained in the Western campaign. France had already had to pay for disregarding the warnings of a Rougeron and forgetting to put a roof on her fortified house. England lost Hong Kong and Singapore despite the clear lessons of the "Battle of Britain". The U.S.A. was taken by surprise at Pearl Harbor and bombed out of Corregidor. Karl Haushofer said quite truly that the art of war had entered a new era. Was the German High Command aware of it? Was it ready and mentally equipped to see the signs of the times?

Shortly after Compiègne Hitler issued the order for the construction of the Atlantic Wall. Vast masses of material, millions of tons of concrete, iron, steel and wood, hundreds of thousands of miles of cable and wire were built in by an army of workers, technicians and engineers. More than a hundred thousand tons of coal and a vast quantity of petrol were needed to transport all this

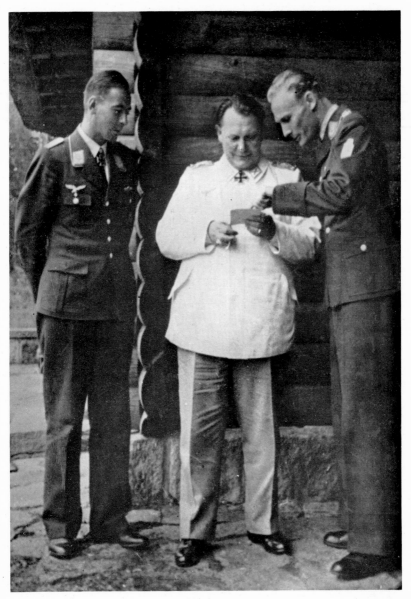
At the Reichsjagerhof Rominten (Major-General Peltz, Göring, the author)

Hitler, in person, laid down the objectives to be achieved during an advance

The author's dive-bombing attack in a JU 88 on the port of Tuapse
in the Caucasus

material and the necessary labour. Meanwhile *grandpère uniformé* sunned himself in French billets and the chair-borne staff flourished in Paris. So it did not sound very convincing when Goebbels uttered fearful threats against the Anglo-American invasion armies concentrating in southern England: "even the attempt is criminal!"

On the 4th June, 1944, the Allies had complete control of the air when the invasion started. After a desperate resistance by weak German forces such as the young volunteer Waffen S.S. divisions in Normandy, Patton's tanks were out in the blue after their break-out at Avranches. The German retreat to the Rhine was a flight which became chaotic under the ceaseless air onslaught of the Allies. There was no halt before the Vosges, the hastily-reconstituted defences of the Maginot Line and the West Wall (Siegfried Line).

At that critical juncture Hitler for the second time entrusted Goebbels with the execution of the measures necessary to continue "total war". The same evening those instructions were issued I travelled with Goebbels in his special train from Führer Headquarters to Berlin. We discussed what should be done first. I was impressed by his brains, his quick and lively intellect, his orderly mind and his common sense. The only wrong note was struck by his frequent displays of vanity.

He knew of my feud with Göring and my last memorandum to Hitler. Neither of us was mealy mouthed. Shortly before we reached Berlin he concluded the discussion with the words:

"All the measures for total war which the Führer handed over to me today are too late. I wanted to get them going two years ago but Göring would have none of it. Even when I closed the luxury restaurants in Berlin only there was nearly a Government crisis. Göring insisted that Horcher should be left open. In my capacity as Gauleiter and Defence Commissioner of Berlin I had it closed. I would have sent in my resignation if the Führer had not personally intervened. Horcher was closed. Look, Herr Baumbach, that sort of thing's going on everywhere! I've been keeping a diary for years. I dictate every morning what happened on the previous day. Perhaps I shall take to authorship at the end of my life. If so, some amazing things will come to light. I wish you good night."

Meanwhile, on Hitler's orders, half East Prussia was shifting earth to build fortifications for the protection of the home soil.

Himmler made one of his rare longer speeches, in which he ranged far back into history in calling the Volkssturm into being. The national heroes of the War of Liberation, Old Fritz and even the mythical Teuton warriors of old, were summoned to witness the fighting resistance of the "whole nation in arms". But despite the revival of the werewolf idea, did anyone seriously believe that T34s, Stalin, and Sherman tanks could be held back by barricades of wood and stone, and Flying Fortresses fought off with sporting rifles?

How did the situation look to Hitler?
Extracts from conversations between the Führer and Colonel-General Jodl, on the 31st July, 1944, at the Wolf's Lair.
Very secret.

Hitler: "When I look today's difficult position in the face, I cannot for the moment see further than the problem of stabilizing the Eastern Front, and I am wondering whether, having regard to the situation as a whole, it is altogether a bad thing that we are, relatively speaking, being squeezed. If we can hold the area we now occupy, it will be sufficient for our needs and we shall no longer have such enormous lines of communication. Of course that assumes that we really give the fighting fronts what we have been able to save from our previous L of C areas. . . . The reduction of the space available to us may not be a disadvantage; it could profit us provided always that we put into the fight every man we have in the gigantic area we still hold. On that assumption it is my firm conviction that even the Eastern Front can be stabilized. . . .

"In my view France presents us with a critical problem. If we lose France as a war area we lose our base for the submarine war. In any event there are a few important war resources we must still get out of that area, the last wolfram supply for instance. Probably the mines could be worked more intensively than hitherto.

"It is equally clear that open warfare there is utterly impossible under present conditions. We can only manoeuvre with part of our forces. We cannot fight open warfare with the rest, not because we have not got command of the air but because it is incapable of manoeuvre. The men are not equipped or armed for it. Our strength in France is not to be measured by the number of divisions we are supposed to have there but by the small number of divisions which are actually mobile. They represent a small fraction. If the area were not so important, the decision would have to be taken here and now to evacuate the coast and immediately withdraw the mobile troops to a line which could be defended in a non-mobile fashion, if I may use the expression.

"But one thing is clear. I have here forces of a certain strength. They can hardly hold their own on this small front. When I remember that by and large seventy-five per cent of all our mobile forces are here and the substantial rest are non-mobile and I imagine them all being transferred to such a line, you can see how impossible it would be for us to hold it—no matter where it is—with what we have available.

"One must be clear that there can be no turn for the better in France until we can regain air superiority, even temporarily. So it is my opinion that, however hard it may be at the moment, we must do everything to ensure that in the last resort we can hold the Luftwaffe formations we are keeping at home as a last reserve in readiness to be employed at some point where we can turn the tables once more. I cannot say here and now where that point will be.

"I regret that it must be a matter of weeks and we cannot act quicker, as to my mind there is no doubt that if we could suddenly pump in an additional 800 fighters and at once bring our fighter strength up to 2,000—as we probably could—the whole crisis now on our hands would be overcome at once. There would be no more crises, though even then it would be true that you cannot fight a war unless you can get some sort of a Luftwaffe back."

During this remarkable talk with the Army Chief of Staff Hitler was still suffering from the effects of the attempt on his life eleven days before. He had something to say about them:

"I would like to have gone to the Western Front, but it is beyond me. For some days I must not fly, because of my ears. If it were merely doubtful whether the other ear is completely cured, I wouldn't care a damn. I would risk it. But if I flew now, what with the noise and the different pressure, the effect might be catastrophic. Where should I be with a sudden inflammation of the middle ear? I must have treatment. My head has not been entirely unaffected either."

Jodl: "Some slight shock to the brain."

Hitler: "I can talk standing for a time, but then I suddenly have to sit down. I would not trust myself today to address a crowd. I could not manage a speech like I made at Obersalzburg recently, because I might suddenly feel faint and collapse. Even when walking there are times when I have to pull myself together not to fall.

"I have to go all over the place in the next few days, so I consider it necessary that the Reichsminister should get over here more frequently so that you don't have to see the Foreign Minister. I would have been glad to discuss everything with him myself but I could only tell him what you can.

"I am sure you are all in yourself too, but for me all sorts of things are always cropping up, many of them vital. In the ordinary way I ought to have gone sick for a couple of weeks or so; but I've had to work at least eight hours a day—and that's not including time spent in reading dispatches.

"But by a bit of a miracle this shock has almost cured my nerve trouble. My left leg always trembled a bit when the conferences went on too long and it used to tremble in bed. Now all that has gone—though I'm not saying that I consider what's happened is the proper treatment."

The important feature about all this is that for the first time Hitler says openly that all his defence measures are bound to fail in the absence of an equally vigorous defence from the air. Fortress Europe too lacked its protective roof.

In the age of air warfare—the Second World War was only a preliminary—all ground defences, even the strongest, are worthless if the defender has not at his disposal a powerful, active air defence in the shape of fighters, highly developed anti-aircraft artillery and a modern radar organization.

Germany could have prepared all this defence system in time, but her war leaders were tied to the conception of land warfare, and until this conversation with Jodl, Hitler, for reasons of prestige, would never admit that the Reich had been forced on to the defensive. Goebbels and Speer told me frankly later that all the measures now taken had been adopted too late.

The German flak artillery had long been inadequate for its titanic task of dealing effectively with the Allied bombers. The increasing speed of the enemy aircraft involved even higher muzzle velocity of the guns. This desirable development was in turn limited by the rate of wear of the barrel. In addition to higher velocity, it was necessary to get better results from hits. An 88-mm. shell could only destroy an aircraft with a direct hit. So in spite of the critical position as regards powder an increase of calibre to 105 mm. and 128 mm. was called for. By 1942 a heavy gun of 240 mm. calibre had been planned. Its barrel life was to be 250 rounds. It had technical novelties, rockets to increase velocity and ground directional aid to improve accuracy. But Hitler and Speer had objections to this project and it was cancelled as early as the summer of 1943. The acute raw material shortage, and not tactical or technical considerations, was primarily responsible for that decision.

Of considerable interest in this connection are the following letters passing between Field-Marshal Keitel and Milch, the Quartermaster-General (Air):

"Dear Field-Marshal Milch,

"On the 5th March Reichsminister Speer had a discussion in the presence of the Führer with representatives of the Armaments Department on the gun programme. An attempt was made to secure some reduction of the flak programme ordered by the Führer in the interests of other types of guns. The Führer has confirmed his order, and all the authorities concerned are strictly enjoined to see that the original flak programme is rigidly adhered to."

Milch made an ironic reply on the 11th March:

"The Luftwaffe accepts the flak programme which has been laid down. The main difficulty is the allocation of copper, as in the first quarter of 1942 the Luftwaffe is receiving only half the copper allocated to it in the last quarter of 1941. The whole of the copper allocation of the Luftwaffe would only cover seventy-five per cent of the flak programme, assuming, of course, that all aircraft production ceases."

The principal culprit in copper consumption was the search-lights, each of which used more copper than a bomber. Although at that time the usefulness of searchlights had very greatly dimin-ished Hitler demanded that their production should be stepped up to the limit.

A month later Milch had this to say on the subject:

"If we are to produce 1,000 searchlights a month there will not be an ounce of copper over for anything else—a state of affairs which is impossible, as we cannot do without aircraft, flak artillery and so forth. We must tell the Chief of the General Staff that the Reichsmarschall has already decided in that sense. The Reichs-marschall knows well enough that the Führer is all for searchlights and will hear of nothing else. He thinks that anyone who does not agree with him is an ignoramus. He inists on searchlights at any cost."

The production of searchlights on the same scale continued even after the Luftwaffe and the navy had renounced any allocation for themselves (owing to the lack of service and maintenance personnel) and more than 1,000 searchlights were lying idle in hangars.

Hitler regarded flak artillery, apart from its usefulness against tanks in ground fighting, as the fount and head of home defence against air attack.

Milch expressed his opinion to the contrary in a remark he made shortly after recommending the urgent development of anti-aircraft rockets in August, 1943: "Flak is not magic and can never make the enemy cease his air attacks. It can force him to high altitudes and cause him losses but cannot keep him away. Only the fighter component can do that."

When the enemy's air superiority made itself felt over the home ground as well, respect for flak revived even among its former opponents. On the 1st August, 1944, Speer said: "In the last few months flak has shown that in the massed attacks on our cities it has brought down more aircraft than was hitherto thought possible. It will become increasingly important. In view of the shortage of aircraft fuel which we must expect we cannot say what are our prospects, both in home defence and in dealing with enemy air forces at the front. But we can at any rate say that flak drives the enemy higher and higher and his accuracy is correspondingly diminished."

At the end of the war flak was practically our only defence against the big enemy raids. Even in November, 1944, Hitler was calling for more and more flak:

"Führer H.Q., 4th November, 1944.

"In his reports on his terror attacks on the Reich the enemy is always speaking of the 'hell' of the German anti-aircraft fire. Many of his sorties have been frustrated by concentrated flak defence.

"To exploit this tactical and psychological factor to the full we must step up the fire power of our flak in every conceivable way. I therefore order the immediate intensification of the flak and munitions programmes. This must extend to heavy, medium and light flak, including their ammunition, ground directional apparatus, sighting and equipment.

"All current experimentation and developments designed to increase the efficiency of the guns and their ammunition, and any other new developments relating to flak defence are to be relentlessly pursued.

Signed: Adolf Hitler."

Among the "new developments" were flak rockets. Rocket development, dating from 1932, had envisaged an unguided powder rocket that could reach an altitude of 23,000 feet and was provided with a clockwork detonator. To the ordinary explosive rockets were added rockets with parachutes and barrage wire hanging from them. With these rockets a barrage could be shot up and held, and renewed, in position over the target until the danger of attack was over. But the development was stopped even before the war by the lack of a feasible self-destroying apparatus and the excessive consumption of powder.

In 1939 the guided flak rocket was again asked for and its production again stopped. In 1941 the General of the Flak took it up again. It was to have a range of twenty-five miles and reach an altitude of ten miles. Detonation could be effected from the ground or by a fuse. It was to be powerful enough to bring down armoured aircraft for certain.

The competent department of the Air Ministry which was

concerned with the problem of guided rockets from aircraft considered that flak rockets were too complicated. Firing rockets from aircraft was much simpler and guaranteed greater accuracy. In November, 1941, it was decided that the development of flak rockets should cease in favour of that of aircraft rockets. Göring and Jeschonnek approved that decision of the Technical Department.

The higher altitudes attained by Allied bombers and their increasing speed and protection against flak splinters compelled the General of the Flak, von Axthelm, to ask for guided flak rockets again in May, 1942. In June Göring approved the following programme: development of a rocket guided visually, and as the next stage an electrically guided rocket which should be self-guided in the vicinity of the target and detonated from a distance. It must be able to frustrate the enemy's evasive action.

In November, 1942, Milch was apparently convinced of the importance of this development. But at first he gave very little help. Differences of opinion about the value of flak rockets still persisted.

Speer and his colleagues, as well as the Luftwaffe leaders, maintained a waiting attitude while Milch cleared away the preliminary difficulties. But we were now in 1944, so that before the war ended the German flak rockets "Butterfly" (radio-controlled), "Waterfall" (based on the V.2), "Rhine Daughter" (radio-controlled), "Gentian" and "Tiger Lily"[1] (for ballistic research) had never got beyond the prototype and test stages. The "Butterfly" seemed the most promising. In addition to unguided rockets from aircraft, a radio-controlled missile made by Henschel, the Hs 298, for firing from aircraft was also undergoing test.

The new invention of radar technique had much the same experience as rockets. The importance of high frequency radio for modern offensive and defensive weapons was insufficiently realized by the German High Command until 1943, though the early years of the war had shown what influence radar technique could have on the course of military operations. We had lost the "Battle of Britain" as a result of a close and well-organized radar network. Our submarine war had been paralysed by the equipping of Allied ships with radar apparatus and the activities of radar patrol aircraft.

The development work of the wireless industry for military purposes was in its infancy before the war, though Germany, like Great Britain, had recognized its importance for military operations. In both countries the spotting of aircraft with the "Brown Tube" had been carried out, and each thought that it alone possessed the secret, or at any rate was well ahead of the other. But Germany, by

[1] Code Names.

restricting the activities of the amateur, had a less solid basis for invention and experiment than England and the U.S.A. where the field was open to amateurs and proved a fertile soil for high frequency technique.

For military purposes an overall organization should have been created at an early stage to co-ordinate invention, development, demand and supply. But this did not happen until the 1st January, 1944, when Göring created a Reichs Inventions Council. Hitherto each branch of the services had worked independently, with no obligation to exchange information.

In the Luftwaffe radar was the affair of Intelligence, as it was in the army and to some extent in the navy. Another impediment was that Milch did not like General Martini, Chief of Air Intelligence, personally. The General, an extremely conscientious and industrious person who had played an outstanding part in the building up of the Intelligence Corps, was not himself enough of an expert to promote and supervise the development of radar, and he was quite unable to cope with all the intrigue around him. To Göring's question on the 2nd May, 1943, as to who was really responsible for the radio industry Milch and Martini had both replied that it was not their concern. Göring decided that it should be Milch's responsibility. But Martini was left to carry on as before.

The scientific side of radar in connection with the Luftwaffe had been entrusted to a commissioner with full powers, Staatsrat Plendl. He paid little attention to any other expert and failed to appreciate what the English were doing. He did not realize that the Western Powers were adopting new methods which defied radar spotting. It became plain that the German scientists were almost always earlier in the field than their Western opponents, but the responsible parties in industry and the Wehrmacht had seldom exploited their own discoveries in time, and often turned down promising novelties in equipment which the enemy was subsequently found to possess. Many devices on the right lines but which could not be immediately put to practical use were rejected as "technically immature".

On the 18th March, 1943, Göring poured out his woes about German radar technique in the following terms: "But worst of all at the moment is the position with regard to our radar. It's enough to drive one mad! We must frankly admit that in this sphere the English and Americans are far ahead of us, worlds ahead! I expected them to be ahead but quite frankly I never thought that they would get so far ahead. I did hope that even if we were behind we should at least be in the race."

In August, 1943, British Bomber Command changed its tactics

and for the first time used new interference and sighting devices in the heavy raids on Hamburg. With a very effective Düppel Interception[1]—throwing out strips of tinfoil, known as "window"—the new "Rotterdam"[2] bomb-sight and a well-rehearsed technique, the English completely surprised the German defences.

The new English bomber stream tactics, which replaced the previous individual approach and retirement and reduced the protracted stay over the target, made the rigid German night fighter tactics no longer applicable. The enemy night intruder operations also had a catastrophic effect on German ground and air radar.

The German radar industry soon produced an answer. The SN-2 "Lichenstein" apparatus would have enabled our night fighters to "see" again. But deliveries to the Luftwaffe were delayed by the Allied raids on Berlin, which seriously damaged the factories concerned. Even in March, 1944, all our night fighters had not been equipped with the new apparatus.

The last phase of the radar war was ushered in by the invasion. In June, 1944, we had the first reports that the Allies were throwing out longer tinfoil strips and our SN-2 was again being foiled. The English raid on Stuttgart was practically without loss. The more skilful use of the Mark XIV bomb-sight, the extremely ingenious radar deception and countermeasure tactics of No. 100 Group of R.A.F. Bomber Command and the system of separate bomber streams made it almost impossible for the Germans to get an overall picture of the attack and follow its course. The feints and ancillary raids of the Mosquito formations added to our difficulties. German night fighters were faced with new problems. The new air radar apparatus FuG 218 and the "Panorama" ground radar apparatus came too late, as the time for producing them in quantity had gone by. The Allies had won the radar war.

This sketchy summary may give some indication of where the weaknesses of Fortress Europe lay, and the mistakes which were responsible for them. Of course I am not forgetting that its foundations had been laid on treacherous politico-idealogical ground which ignored the real strategic and technical problems. All the Allied war planning had to do was to exploit these weaknesses. The proper result was obtained by amphibious landings under air cover, with mighty bomb carpets on the advanced defences and systematic bombing of the transport network, industrial installations and centres of population.

[1] We had carried out similar experiments at the Düppel estate near Berlin. Hence the German name. Göring stopped them.

[2] It first came into our possession when an English aircraft crashed near Rotterdam; this was the Mark XIV bomb-sight.

Experienced leaders in air warfare and technical experts gave us prompt and repeated warnings of the danger threatening us, but no or insufficient attention was paid to them. The men at the top remained obsessed with the idea that a decision could be obtained only by offensive strategy—whether on the Volga or in the Urals. They insisted on the production of heavy tanks, bombers and other offensive weapons long after it was clear that the emphasis in armament must be placed on defensive equipment. So Fortress Europe remained a utopian conception in the absence of the technical material required to maintain it. It was a serious offence against the German nation, and the finest achievements of our fighting men were unable to atone for it.

CHAPTER XV

THE BOMBS FALL

THE pioneer of the air war against Germany was the Royal Air Force. The year 1940 witnessed the first daylight attacks on industrial targets in Germany. They were abandoned because losses were too high. Then came smaller scale night attacks on oil refineries, aluminium and aircraft factories. These too were abandoned. With the technical means available at that time it was not possible to hit the targets. Hence attacks against city and industrial areas.

In conformity with the experience gained in the First World War the defence system in Germany was prepared as the protection of particular targets. For the defence of vulnerable industrial or military installations light fighter squadrons and a few night fighters had been provided. Ground defences comprised heavy and light A.A. guns, searchlights and aerial barrages. In view of the particular vulnerability of the industrial areas in west and south-west Germany and the limited range of foreign bombers before the war, "Air Defence Zone West" was comprised in the Siegfried Line system.

The German High Command believed that with this zone fully manned, enemy aircraft flying at less than 23,000 feet would be exposed for three or four minutes to a concentrated, gap-less barrage: "The enemy over the zone will suffer continuous losses and have his effectiveness steadily drained away, particularly as he must pass through it twice, out and home. If he wants to get out of range he will have to ascend to a vast height at which accurate bombing will be hampered by poor visibility."

Up to the autumn of 1939, 197 battery positions for heavy flak and 48 for light A.A. had been built in the Air Defence Zone West. Shortage of material had prevented the supply of searchlights before the outbreak of war, but it was then decided to create a continuous searchlight belt. No other country had provided anti-aircraft defence on its frontiers on a similar scale.

The first enemy raids avoided the A.D.Z.W. however as they took the sea route. On the 4th September the German fighters obtained a surprising defensive success when the R.A.F. sent Blenheim bombers without fighter escort to attack the naval station of Wilhelmshaven. The enemy was caught by the "Freya" apparatus, which had been installed as an experiment and got the fighters into the air in time.[1] Unfortunately, we exploited our victory in another air battle over Heligoland on 26th September solely for propaganda purposes, whereas it had the most important air-tactical lessons for both attacker and defender.[2] Göring became so confident that he said that he would eat his hat if a single enemy aircraft appeared over Berlin.

When the British Bomber Command decided at the beginning of 1940 to go over to night raids, we at first tried to fight them with Me 109s and searchlights. The results of this improvisation, especially on account of its dependence on weather, were unsatisfactory, so a Night Fighter Group of two-engined Me 110s and Ju 88s was formed. Until the end of the year this night fighter group formed part of a night fighter division which was under the command of General Kammhuber. Its strategic task was to assure command of the air over the Reich and the occupied territories by night. Yet the development of our night fighter force and the methods it adopted suffered both from its local association with the units assigned to the protection of particular targets and the rigidity and formality of the ideas cherished by our air leaders. They could not adapt themselves to the ever-changing flying and attack technique of the British. The business of operating a few aircraft in regionally limited fighter areas hindered a more sustained pursuit and battle with the invading enemy bombers.

The first year of the war was enough to show that the material we had prepared and the fighting tactics we employed were inadequate for successful defence against an enemy air offensive. But we had to make do with what there was, as further material for the development of our air defences was withheld. On the contrary, flying and flak units were withdrawn to support the army at the front.

The R.A.F. delivered its first great blow against Germany on the night of the 30th May, 1942, when Cologne was attacked by 1,000 bombers. Forty bombers were brought down. A year passed before the city was the target of another heavy attack. The lovely Altstadt was laid in ruins and the cathedral was seriously damaged. On the 5th March, 1943, there was a raid on Essen in almost the same strength and with the same terrible effect. These were the opening moves of the great air offensive against Germany. To the Allies it was the revelation of German weakness at the most sensitive spot and it should have been a last minute warning to our High Command.

[1] Seven out of twenty-nine Blenheims were lost.
[2] Five out of eleven Hampdens were lost.

At the Casablanca Conference the two great Western Allies decided on the ways and means for a victorious conclusion to the war. Total victory must follow a total war. It would be brought about mainly through a strategic air offensive, which would aim primarily at destroying military and industrial installations and the transport network, breaking down the morale of the civil population by terrorization from the air and paralysing our submarine war.

When this last objective of Allied air strategy had already been attained in the first half of 1943 there was a revision of the Casablanca plan at the beginning of June and a new decision was taken —the main weight of the Allied bomber offensive should now fall on German aircraft and ball-bearing production. Even greater emphasis was laid on the continuation of the terror attacks as, after the effects obtained in Italy, it was hoped that Germany's resistance also could best be broken down within the walls of their cities. After the systematic air attacks on West German cities, on the 24th July, 1943, the R.A.F. opened its first great raid on Hamburg. Others followed in rapid succession.

I was an involuntary witness of this frightful catastrophe. Approximately 60,000 dead lay buried in the ruins of Hamburg. The city had become the Stalingrad of the German homeland, an ominous warning of approaching disaster.

The Americans did not lag behind. They surprised us by coming from the South. On the 13th August U.S.A.F. bombers from their North African bases made an attack on the Messerschmitt works at Wiener-Neustadt. This was the prelude to a series of heavy onslaughts on the aircraft production industry and the creation of a second air front which forced on us a more ample decentralization of our air defence and thus the dispersion of our already inadequate forces. Four days later, on the 17th August, more than 200 B-17 "Flying Fortresses" attacked Schweinfurt, the centre of our ball-bearing production, and dropped more than 1,200 tons of bombs. The enemy lost thirty-six aircraft and our ball-bearing output was reduced by thirty-five per cent. On the 15th October a second attack on Schweinfurt met a solid German defence. Of the 228 enemy bombers engaged sixty-two were shot down by fighters and flak and a further 138 were heavily damaged or rendered unusable. In the official U.S. report *Strategic Bombing Survey* of 30th September, 1945, it is stated that this was one of the most decisive battles of the war; further losses on that scale could not be borne; deep penetration into Germany without fighter protection had to be abandoned and attacks on Schweinfurt could not be renewed for four months.

In that period our ball-bearing production could be brought up to its old level. The heavy losses in the attack on Schweinfurt forced the enemy to adopt very different tactics in daylight bombing in the Reich. Hitherto they had believed that the heavily-armed bomber forces, flying in tight formation, could penetrate deep into the country. Now they were dependent on an escort of long-range fighters—a technique the U.S.A. had developed for even longer distances in the Pacific. In December, 1943, escort fighters of the P-51 Mustang and P-47 Thunderbolt types were available for the European theatre. Under their protection there was a notable revival of the American daylight raids. Though at first these escort fighters confined themselves mainly to the protection of the bomber formations, they soon used their increasing numerical superiority to challenge the German fighters or seek out suitable targets for themselves. German fighter losses mounted at an alarming rate. In January and February, 1944 respectively, 1,115 and 1,217 of our fighter pilots were shot down, among them many senior and experienced leaders. By the spring of 1944 the resistance of the German fighter defence had been broken. We could put only eighty fighters into the air to cope with the great attacks of the 12th and 28th May on the transport networks of West Germany and France, which were carried out by all the aircraft available in England as a preliminary to the invasion.

Meanwhile, in 1943 the R.A.F. had systematically stepped up their nightly attacks on Berlin and almost all the other big cities in the West. After the Hamburg catastrophe the inhabitants of such cities were all left wondering when it would be their turn. Even though they showed their discipline and, apparently unmoved, took to the shelters and newly-built bunkers when the sirens (which native humour christened "Meiers Hunting Horn") sounded, there was no hiding the nervous tension prevalent in the capital. My wife's diary is eloquent on the subject.

When a spell of bad weather just before Christmas gave Berliners a short respite, Goebbels announced with relief: "that was balm on the capital's wounds." Yet even while Berlin was recuperating, the bombs rained down mercilessly on Munich, Nüremberg and Stuttgart. Even big cities further east such as Stettin, Danzig and Königsberg fell victims to the British bombs. It was not until the late summer of 1944 that the R.A.F. began to pick out industrial targets suitable for night raids. The air terror against the German civil population had not been able to attain its object, to break down morale. It had not even succeeded in paralysing our industrial production, if only because the most important armament factories were away from towns and the drop in output resulting from the air attacks could be made good by overtime.

The behaviour of the German people under the enemy's air terror was worthy of all praise and Goebbels was not wrong when he described it, with his particular brand of pathos, as "almost religious".

When the fighting fronts drew closer to our frontiers the target area became even smaller. There was no escape anywhere. It was in vain that the German leaders now tried to create a mammoth fighter force to recover command of the air over the Reich. It was too late. Our production possibilities had been destroyed by the English and Americans. Then the German transport network was systematically destroyed. It was not long before not a horse-drawn vehicle, let alone a lorry or train, dared to show itself in the daytime.

With its lack of suitable aircraft and crews, the Russian long-range Aviation Command, the ADD, did not participate in the strategic air war against the Reich but confined itself to the rôle of co-operation with the Red Army. Up to the middle of April, 1945, British and American bombers made "Fortress Berlin" ripe for assault while laying bomb carpets ahead of their own armies pressing across the Rhine. The "steam-roller" tactics of the Russian tank armies in the East were the equivalent of the Anglo-American bombers in the West. Much of the bombing was without any visible military necessity. Dresden, crammed with refugees, was the target of a night attack even more terrible than the Hamburg catastrophe. In Leipzig, the centre of the German publishing industry, apart from many thousands of human victims approximately two million books and irreplaceable printing material were destroyed. Seventeen thousand corpses lay under the ruins of Heilbronn, a mediaeval town full of timber-framed houses. Würzburg, a gem of German baroque, was reduced to rubble and ashes by the advancing Americans.

Shortly after the capitulation, in a report made by an American Commission to Congress on the effects of the strategic air bombardment of Germany, the following figures were given:

In Germany, twenty per cent of all dwelling-houses—more than 3,600,000—destroyed; 300,000 civilians killed, 710,000 injured; Germany's big towns turned into hollow walls and rubble.

I cannot find words to describe the scenes of horror in that terrible inferno. Of course it is part of war itself that many persons lose their lives and the fate of individuals is lost sight of and becomes unimportant. But the fact that in those nights of bombing the German people did not sink into universal chaos and hopeless nihilism, and that the forces of revival are now at work, is a proof of its unbroken will to preserve its faith and survive.

CHAPTER XVI

WHERE WERE OUR FIGHTERS?

WHERE were our fighters all this time? This question was always being asked when we pilots were home on leave. The nation was prepared for anything and would face even heavier sacrifices. Even the defeats in the East seemed less depressing than the Anglo-American air terror growing month by month. An uncontrollable fury possessed even the most reasonable people when they saw the German sky filled with American and British bomber formations flying as if at a review, and when they stood by while their houses and cities—all they had left—and their relatives fell victims to the falling bombs without any reply to this terror. Where were our fighters? Where were the victorious and much decorated heroes who had been the national darlings in the first two years of the war?

The German fighter arm had entered the war with an aircraft, the Me 109, superior to all enemy machines. It had proved itself in Poland, the West and the duels fought over and beyond the English Channel. Yet this aircraft had shown itself unsuitable for defence by reason of its short flying time and navigational difficulties in bad weather. The technical development programme had been particularly unfavourable to the fighter arm. Urgent demands for faster types and more powerful engines were rejected again and again. So at the beginning of 1942, the Me 109, with a few slight improvements, was still the standard fighter.

Our fighter production had fallen behind that of the enemy powers and the question now arose whether we should accept that state of things as a fact, or either create a series of the fastest and most modern fighters or first reach parity with a rapid maximum production of a series of old types. The German High Command decided in favour of the last alternative and tried to win the production race with mass production of the Me 109 and Fw 190. But in spite of rising output and continuous improvement of these types, we did not succeed in restoring the technical superiority of these German fighters. Göring admitted it frankly on the 12th September, 1942:

"By and large we have had a certain superiority and an unquestionable superiority at first, so far as fighters are concerned—the Me 109 in its various developments and then the Fw 190. Both types have been caught up with and to some extent overtaken by the

The author gets a direct hit on a big Russian tanker in Sevastopol harbour

Field Marshal Erhard Milch with the author at Headquarters

Colonel-General Jeschonnek in conversation with the author

English and American fighters, particularly as regards climbing powers. To my annoyance they also seem to have a greater range, through using additional tanks which last longer than ours, and this is very unpleasant. Above all, the Spitfire is ahead, a thing our fighter pilots don't like."

The view prevailing in the Technical Department that it was possible to get better performances out of types otherwise unchanged by giving them more powerful engines was now tried on the Me 109 also. At the outbreak of war this aircraft was equipped with a 1,100-h.p. DB 601A. The results were the same as had been already revealed in the case of bombers. Thanks to additional equipment, more armour and armament, there was little improvement in speed and rate of climb. More powerful engines and the increased fuel consumption they involved necessarily reduced their flying time. From $1\frac{3}{4}$ hours at the beginning of the war it fell to $1\frac{1}{4}$.

From 1942 onwards the tactics of the Allied bombers and fighters indicated that they would fly into the Reich at ever higher altitudes. That called for the raising of the maximum altitude of our fighters to 35,000 feet by day and 25,500 feet at night.

Such altitudes were beyond the early models of the Me 109 and the Fw 190 as normally equipped, and the latter could only reach them lightly armed. Both aircraft spent half their flying time in attaining the required altitude. At such altitudes they were inferior to their opponents.

The Quartermaster-General (Air) and the Inspector of the Fighters believed at that time that it was impossible to do without a large number of fighters. They demanded:

1. The production of normal fighters, i.e. a high output of the already proved standard fighters.

2. Performance fighters. These were to represent the acme of technical progress at any given moment.

Although the Me 109 was technically outmoded and equal only to its opponents in the East, large scale production was provided for in the 1944–45 armaments programme.

First among the new aircraft which were to repair the defects of the German fighters was the Ta 152, a product of the brain of Professor Tank, head of the Focke Wulf Aircraft concern. It was to replace the Fw 190. But the first trials showed that with existing engines it was not materially faster than its predecessor. The Inspector of Fighters had asked for centrally-placed armament and this development was carried out on this aircraft. The irresolute attitude of Milch and the failure to supply the 2,000-h.p. engine anticipated, delayed serial production until the autumn of 1944. Before the war ended only 67 Ta 152 fighters had been produced, so that they were practically non-existent. With the DB 603LA engine the Ta 152 was to have a flying time of three hours. The Me 209 and the Me 309, projected replacements for the Messerschmitt Me 109, were technical failures at first, like the earlier destroyer fighter Me 210 and the Me 410. Only a few prototypes were produced.

The Dornier Do 335 might have brought some relief as a new type. In the words of Milch (17th December, 1943): "It should hold its own in speed and altitude with the Lockheed P-38 Lightning and does not suffer from the unreliability of its power units." But before the war ended only eleven Do 335s had been delivered.

Even in respect of flying personnel the German fighter arm had fallen behind. The continuous high losses had so lowered the quality of the crews that only a few experts were capable of successes against our Western opponents. Our great successes in Russia could not blind us to that fact.

The poor flying-time and lower speed led to frequent failures and rising losses. Inadequate armament limited the prospect of success in shooting down opponents and so had a prejudicial effect on the offensive spirit. An insufficient fuel supply dictated the moment for engagement without regard to the tactical situation in which the fighter pilot found himself. All too often he had to break off his attack before he was really in contact with his adversary.

While the Anglo-Americans could carry out attacks in bad weather, and with the ground invisible, the German fighter pilots had no equally effective radar instruments, nor were they in a position to ward off these day and night raids. Just as the Fighter arm in the building-up stage had been weakened over and over again by the division of its squadrons and the surrender of personnel to other formations, the shortage of fuel for training purposes now made itself felt in no uncertain fashion. Icing troubles and the absence of the pressure cabin required at high altitudes substantially diminished the fighting value of the German fighter. Worst of all, the reliability of the engines fell off to such a degree that from the summer of 1944 onwards our fighter pilots lost confidence not so much in themselves as the aircraft they were flying.

The position was little better with regard to night fighting. At the start the most suitable aircraft for that purpose (besides the Ju 88) was the Bf 110, which had not particularly distinguished itself as a destroyer and strategic fighter. It had a useful rate of climb and sufficient manœuvrability. But its flying-time was low. On occasions of deep enemy penetration into the home area the Bf 110s usually had to break off attack and pursuit prematurely through shortage of fuel. After the Ju 88 the Do 217 had considerable success, particularly in pursuit at night, thanks to its rate of climb and long flying-time.

On the 27th June, 1943 Major Hermann again pressed on Göring his plan for night pursuit with single-engined fighters flying in areas lit up by searchlights. In his view the searchlights created daylight conditions over the target to be protected so that the ordinary day fighter could suddenly emerge from the darkness and approach and shoot down at close range its victim caught in the beams. Moreover it would be possible to direct the fighters engaged by radio.

Major Hermann himself gave many successful demonstrations of this technique during the big enemy raids into the Ruhr and over Berlin. But there were also many occasions on which he had to land by parachute and not in his aircraft. His squadrons had been hastily got together and after many had been shot down the losses rose so steeply that his method of night fighting by daylight, familiarly known as "Wild Boar", was abandoned. The English also began to carry out raids in bad weather, and even though they could not see the ground and this automatically made them safe from the German single-engined night pursuers. In the long run, the difficulties of night pursuit with such aircraft and methods of direction could not be overcome. Göring had this to say about it: "It is not right that I should always have to be drawing on the bombers. Both day fighters and bombers are not one hundred per cent suitable as night fighters." Even the few raids of the Russians, which in 1942 got as far as Berlin at altitudes of between 23,000 and 26,000 feet, were not effectively dealt with by night fighters.

In addition to having a very high rate of climb the night fighter had to be at least as fast as the fastest enemy bomber, and further it was a condition precedent that it should be effectively directed from the ground. But the stage reached by our radio and the building-up of our night fighting organization had made anything but rigid or improvised directional methods impossible. They usually only gave the night fighters one chance in a burst of firing. The moment the enemy bomber made a violent swerve or changed its altitude only six or nine hundred feet, all the calculations were thrown out and the night fighter seldom or never found his adversary again—and then only on bright moonlight nights. Yet their losses were less than those in day fighting. At the end of the war modern night fighting methods were no longer fully effective.

It was only in the second half of the war that two important technical developments matured; jet and rocket propulsion.

In Germany a large number of technicians had been engaged in evolving this new form of propulsion, which meant enormously-increased performance, for various special branches of aviation, the artillery and the navy. Manned and non-manned missiles quickly attained speeds exceeding that of sound, new gunnery records were achieved by rocket weapons and an unsuspected use for them was found in the field of U-boat and torpedo technique.

The first practical experiments with jet aircraft were made in Germany in 1937–38. As opposed to the aircraft engine which propels the aircraft with the help of airscrews, jet propulsion gets its motive power from the reaction of the hot gases streaming through the exhaust at high speed. The three essential elements of a jet engine are the blower, the combustion chamber and the turbine with its exhaust. It has fewer individual parts than a piston engine and its operative weight is about half. For fast flying it has great advantages, and jet propulsion makes better working conditions and simpler prototypes possible. But the fuel consumption is higher.

The German motor industry had reached the conclusion that there was a limit to the possibility of improving the performance of the petrol engine, a limit which could only be raised very gradually, if at all, in view of the raw materials available. So a solution of this problem had become an obvious necessity. Research and scientific work in the field of jet propulsion dated from the First World War, though neither then nor in the post-war period were they systematically followed up. It was the same story with rocket propulsion.

Hitherto powder rockets were used to carry small things such as Very lights and life-lines for short distances. A greater use of the rocket principle, which was worked out by Oberth and de Valier, among others, did not at first lead to the goal. De Valier lost his life in a racing car in a rocket experiment carried out in conjunction with Fritz von Opel. The idea of using that form of propulsion for aircraft was then taken up in various scientific circles.

In rocket propulsion the engine has to take with it the supply of oxygen necessary for combustion, while in jet propulsion, as with the normal petrol engine, the oxygen is obtained from the surrounding air. There is no additional oxygen and its container.

The principle of jet propulsion is based on the following process. A turbo-blower compresses the air sucked in to about four atmospheres and, after it has been heated by the compression, forces it into one or more combustion chambers. Oxygen is continually fed into these combustion chambers and a combustible mixture produced. This is ignited, resulting in hot gases which drive the turbines and then stream through the exhaust at high speed. The reaction of these gases creates the effective thrust which propels the aircraft.

Turbine jet propulsion also represented an important factor in the oil situation in Germany. Lower octane fuel, hitherto suitable only for heating purposes, could be used. It had extraordinary advantages for aircraft bodywork also. In the absence of the heavy petrol engines and bulky airscrews, the whole construction could be better designed aerodynamically. In high-speed aircraft the useful load was bigger. An appreciable advantage was that it required less manpower to produce a jet than a petrol aircraft.

None the less it proved a long and hard road, with many set-backs, to our first reliable jet. There were many new problems to be solved in connection with the combustion chambers and the control of temperatures rising to 700 degrees Centigrade in the turbines. Moreover, behaviour at high altitudes and starting and stopping technique were unknown quantities from the technical point of view and could only be ascertained from the first trials.

In 1939 the Technical Department had called for a jet aircraft with the following specification: speed 575 m.p.h., thrust 13,200 lb., reducing with the altitude but so that at 30,000 feet altitude at least fifteen per cent remained.

The Air Ministry did not consider this new development as of decisive importance. The August, 1940 ban on all aircraft design the development of which would take more than six months also applied to jet aircraft, notably the Me 262, which would have been the first to be ready for serial production. In addition to Messerschmitt, the Arado concern had produced a jet aircraft, the Ar 234. Despite the ban, these firms carried on with these projects.

The views of the soldiers and designers can best be gathered from their own words at a great conference in Quartermaster-General Milch's presence at the beginning of 1942, when much valuable time had already been lost.

First let us hear what Franke[1], the Director of Testing at the Luftwaffe Testing Ground at Rechlin, had to say:

"The most important thing to remember is that we must expect to find our opponents appearing with jet propulsion. If they do, and we have nothing to counter it, it will be a very serious blow to us. We must not give up jet propulsion whatever we do, whether or not a fast bomber with petrol engines would give us the quickest answer."

Professor Messerschmitt, the designer of the Me 262, answered him:

[1] Quite against his will, Franke had been hailed as 'the man who sank the English aircraft-carrier *Ark Royal*'.

"I agree that the speediest answer calls for a petrol engine aircraft. But we must also produce jet propulsion aircraft, based on our existing jet propulsion equipment. Such aircraft will very soon be decidedly faster than aircraft with airscrews. Whether they can be used in the near future is another question, for if the enemy has nothing of the kind himself it might in some circumstances be a mistake for us to employ jet bombers. Out of a hundred or so we might then possess, one might fall into enemy hands and he will be copying it in no time."

He was followed by Galland, one of the most successful fighter-pilots against the Anglo-American fighters:

"We have now reached the point where the serious position of our fighter defence makes us call for an aircraft with at least equal performance to enable us to carry on; if possible one with superior performance to outmatch them. We do not know what the enemy is doing in the whole field of aircraft construction, but I consider that we shall be making the greatest mistake if we take too narrow a view in our planning and specify for a fast bomber without any account of the fact that the same aircraft, or some similar aircraft, may in an emergency have to be used as a fast fighter rather than a fast bomber. I think that the Technical Department should specify for fighters as well.

"My second point is that our development of jets may well start something of which we cannot see the end. Under some circumstances the use of these things can anticipate events and lead to our giving away a weapon before we possess it in really effective quantity. But I believe we need not be afraid of our own courage—assuming that we go all out. But I have the impression that we do not."

Messerschmitt commented on Galland's view:

"I am of the same opinion. Our fast bombers will have to have petrol engines. But we must simultaneously produce at least a hundred jet fighters in case the enemy turns up with them. The aircraft industry can do it. It has the capacity to turn out jet fighters which could also be used as bombers."

Heinkel, the head of the well-known German aircraft concern, gave his opinion:

"Having regard to our existing jet production facilities I am convinced that in a very short time we could have jet bombers, and jet fighters nearly as quickly. I cannot understand why there should be a moment's delay. We must get down straight away to jet bomber production, on the lines that we, and Junkers too, have laid down. It is fairly feasible to have the jet bomber and jet fighter in one. We have got 575 m.p.h. in any event.

I can take away one of the tanks forward and replace it with six cannon. Or we can place the six cannon—as we actually have done—immediately over the engines. In that way you kill two birds with one stone. If we make a start today we can produce the fighters and the fast bombers at the same time. I think we must go all out to have fifty or sixty of these bombers or 575 m.p.h. fighters by 1944. It is possible. But a beginning must be made. There must be a commissioner of the Ministry to clear away the obstacles. Whether we use Junkers or BMW or any other engine, does not matter. The aircraft can be produced by the middle of next year for certain. Fifty, sixty, a hundred aircraft with a speed of 575 m.p.h. The enemy can be along with it any day."

Franke's final comment at the end of this remarkable conference:

"We must not forget that we know all about the old piston engine and its airscrews, and quite enough about its drawbacks, while we are still a bit in the dark about jet propulsion."

It was natural that Messerschmitt and Heinkel wanted to see their plans and ideas materialize. They were essentially designers. As so often before, there was a lack of decided ideas on the part of the military. Galland's views, based on active service, were nearest the mark. He had had personal experience of the growing superiority of the English and American fighters. He was to renew his demand for a strong fighter force over and over again. It was only in 1944, when it was too late, that the Fighter Staff got what was required.

It is necessary to state that after the discussion at this conference it was decided that the jet plane should be a bomber, though its immediate adaptation to a fighter was to be possible and its construction so provided.

A year later, on the 28th March, the spokesman for jet propulsion had this to say: "The Me 262 is now in existence with a prototype at Messerschmitt's with Junkers engine. They have got well ahead with the tests. According to Messerschmitt, they have been flying at a speed of 530 m.p.h."

The Messerschmitt concern had inaugurated a so-called master plan; forty aircraft should be ready by the end of 1944 provided that the technicians working on the project should not be taken off for anything else and they were given some additional men.

The figure of forty aircraft for the end of 1944 clearly shows how much importance the authorities attached to this project. Messerschmitt was given the extra labour and allowed to get on with the job without interference, but no one bothered about giving this aircraft any priority over piston-engined bomber and fighter production. Even Field-Marshal Milch did not assert his authority in favour of jet production. Neither he nor his colleagues showed their hands or took special action, as Heinkel had proposed.

It was only when Galland flew the Me 262 on the 22nd May, 1943 that the Technical Department did more for that aircraft. He reported to Milch: "The Me 262 represents a very great advance which will assure us an inconceivable advantage if the enemy sticks to piston engines much longer. From the flying point of view the fuselage makes a very good impression. The engines function perfectly except at take-off and landing. The aircraft offers completely new tactical possibilities."

The Air Ministry, Waffen S.S. generals, General Staff and Göring then produced an officer with outstanding technical accomplishments and unusual service experience in German altitude tests and special weapons. He was Major Knemeyer, subsequently Chief of Aircraft Development in the Technical Department. Although at first he had no special position, he was able to convince many leading men in the Luftwaffe of the urgent necessity for jet aircraft. It was really thanks to him that the new technique of jet and rocket propulsion materialized.

In the summer of 1943 the General of the Bombers offered on his own initiative to forgo a considerable number of bombers in favour of jet fighters. In a conference in the presence of the Quartermaster-General on the 18th August, I said to Field-Marshal Milch:

"As home defence must be made strong first—I say that as a bomber officer, although anyone can appreciate that it must go against the grain—I propose that in the present situation we give up the He 177 and its capacity in favour of the jet fighter and jet propulsion."

This offer was countered with the argument that to forgo bombers would not help because quite a different sort of industrial capacity was required for jets, and that in any event there could be no question of stopping the production of the He 177. Bombers were needed more than anything else in the immediate future.

That was the decision of the high-ups while the first heavy raids on German cities had already started and it was certain that we were facing an Allied bomber offensive. Schwenke, the Engineer Director on Field-Marshal Milch's staff and the man responsible for collecting all information from abroad about enemy armament, knew all about the preparations for a strategic onslaught on German industries. He reported every week to his Chief in the presence of several representatives of the Air Ministry, so no one can say that he did not know. But the high-ups simply *refused* to believe what he told them.

By the autumn of 1943 opinion had changed sufficiently for production of the Me 262, as the first jet fighter, to be pushed forward. The question whether the old Me 109 should be dropped was discussed, and then it was decided that the Fighter arm should be given jets as well. Milch in particular was averse to abandoning the old fighter types, the production of which was actually then being stepped up, though they had long been out of date in service and could never bring us any real relief at the front. Our losses with these types were mounting from month to month.

Towards the end of 1943, the C.-in-C. of the Luftwaffe—convinced by his "stargazer" Knemeyer—took a personal interest in the Me 262 and Milch suddenly changed his mind. But even then he would not accept responsibility, as appears from a conference on the 12th November, 1943: "The only thing to which we cannot give a hundred per cent reply is the question whether the Me 262 and jet propulsion are now so foolproof that we need have no hesitation in adopting them next year—and not merely as regards development but from the manufacturing angle also?"

General Vorwald, Chief of Staff of the Technical Department: "After the last conference, yes!"

Colonel Petersen commanding the test sites: "We must always expect set-backs!"

Meanwhile Hitler had intervened and pronounced that the Me 262 must be produced solely as a bomber.

In addition to jets so much progress had been made with rocket propulsion that a new prototype, the Me 163, could be tried out.

Meanwhile the Allied bomber offensive had been unleashed in full fury. The plight of the German fighter defence became more catastrophic from week to week. At the end of April, 1944, the General of the Fighters made no bones about it when he renewed his demand that the Me 262 should be a fighter:

"The problem which the Americans have set the Fighter arm is— I am speaking solely of daytime—quite simply the problem of superiority in the air. As things are now, it is almost the same thing as command of the air. The ratio between the two sides in day fighting at the present time is between 1:6 and 1:8. The enemy's proficiency in action is extraordinarily high and the technical accomplishment of his aircraft is so outstanding that all we can say is—something has got to be done! In the last four months we have lost well over a thousand men in the daytime. Of course that figure included many of our best group, wing and squadron leaders. I have mentioned this last possibility in many a report and conference, and gone so far as to talk of the danger of collapse. Now we have reached the point, because the numerical superiority of the enemy has become so great, that we must ask ourselves whether the fight is not becoming extraordinarily unprofitable to us.

"What must we do to change all this?

"The first thing is to alter the ratio. That means that industry shall produce aircraft in numbers (on which we can absolutely rely) which will enable us to build up the Fighter arm. Secondly, just because we are numerically inferior, and always will be—let there be no doubt about that—technical performance must be raised. It is obvious that for the defence of the Reich we must have the 2,000 h.p. engine. I am absolutely convinced that we can do wonders even with a small number of greatly superior aircraft like the Me 262 or the Me 163. For the fight between the fighters—which in day-time is the preliminary to going for the bombers—is largely a matter of morale. We must break the enemy's morale. With the help of these two factors, numbers and performance, the fighting value of our formations, and the level of their training, will inevitably be raised. I do not expect that we shall ever be on equal terms with our opponents but we shall have a reasonable situation once more.

"In the last ten battles we have lost on an average more than fifty aircraft and about forty men—that is five hundred aircraft and four hundred airmen in ten great raids. At the present pace they cannot be replaced with others equally well-trained.

"I have one other request, and it is that the extraordinarily fine efforts to keep up our numbers should be supplemented by clear realization of the fact that the performance of our aircraft is equally important. We need higher performance to give our own fighter force a feeling of superiority even when we are much inferior in numbers. I can sum up my own feelings in a few words: at the moment I prefer one Me 262 to five Me 109s."

So much for Galland. He now came out quite openly for the high performance aircraft, the Me 262, and rejected the view of Milch and Saur, the head of the Fighter Staff, that we should confine ourselves to producing the largest possible quantity of the older types. Those few words are eloquent of the plight of the German fighter defence at that time.

Hitler was not to be moved from his decision that the jet aircraft was to be used as a "blitz bomber". For quite a considerable time no one dared mention the Me 262, much less its use as a fighter, in his presence. But the fighter leaders and Colonel Knemeyer, the new head of aircraft production, continued to call for the Me

262, so in June, 1944, Saur got the following decision from Hitler (the words are Saur's own):

"The production of the Me 262 as a bomber is dictated by the present state of the war. But at the moment when, firstly, it is available in sufficient numbers for bomber purposes and, secondly, the Ar 234 is available as the jet bomber, the Me 262 can ultimately be a fighter. The Führer has no objection to our minuting this aircraft in the construction programme as for 500 fighters and 500 bombers."

Herr Saur was quite right to talk about "minuting". At a time when the German industry was sorely stricken he was already speaking in figures which existed only on paper. He did an enormous amount of damage with these wild figures, bluffing Hitler over and over again, and deceiving him into thinking that we had enough aircraft of the best quality. There can be no doubt that he had done wonders in the production of the old types, the worn-out veterans, but he eventually fell victim to his own mania when, at the beginning of 1945, his numerical house of cards collapsed in a few weeks.

In the late summer of 1944 Hitler was pressed on all sides to withdraw his order and release the Me 262 as a fighter. After many conversations which I had with the Reichsführer S.S., both Himmler and Speer seem to have supported us. But it was only towards the end of 1944, just before the Ardennes offensive, that Hitler let himself be convinced. Now the Me 262, first a fighter, then a bomber, must be reconverted to a fighter! It was too late! Its old defects and failings reappeared, and had to be remedied at the front.

The delays over the Me 262 were responsible for another project. The Production Section of the Technical Department conceived the idea of a cheap jet aircraft which could be mass produced and put into service very quickly. At the beginning of 1944 a specification was issued to four or five aircraft factories.

Professor Heinkel thought that his He 162 met the requirements. In the summer of 1944 the decisive conference was held in Göring's presence. After violent disputes between Saur and Galland, the He 162 was accepted and given the name of "Volksjäger" (People's Fighter). Messerschmitt and Tank were strongly opposed to this project. They proposed the alternative that the manufacturing capacity for the Me 262 should be increased. Saur remained obsessed by "his idea". He was already seeing the first batch of his Volksjägers rolling from the presses and freeing Germany from the nightmare of the Anglo-American terror.

In the winter of 1944 he had been in contact with the Reich

Youth leaders and the N.S.F.K. (National Socialist Flying Corps). The results of his talks with the N.S.F.K. were as follows:

"Colonel-General Keller, Corps Commander of the N.S.F.K., places all the men and factories of the N.S.F.K. at the disposal of the He 162 project and agrees with Saur's views. It is intended that a whole year's intake of the Hitler Youth shall be assigned to the He 162 without preliminary training on piston-engined aircraft. Suitable gunnery training shall be given on the ground."

With the help of Artur Axmann, the sensible Reichs Youth Leader, and his Staff Leader, Moeckel, who was killed in a motor accident shortly afterwards, I managed to convince nearly everyone of the absurdity of this idea. Even the He 162 called for thorough fighter training and could not "be flown by any Hitler youth". But Saur continued to advocate his project with might and main, and no technical or other argument could convince him that the "Volksjäger" was no "people's fighter". The first production models were ready in February, 1945, but apart from a few non-operational flights it was soon forgotten.

Towards the end of 1944 the first Me 262 squadron, under that brilliant holder of the diamonds, Captain Novotny, came into action. Galland, who had been demoted to Inspector of Fighters by Göring (a murky business!) took over this formation, which in November, 1944, was enlarged to a wing under the command of an experienced fighter expert, Steinhoff. In the last months of the war fighter sorties were carried out from Brandenburg-Briest and München-Riem which clearly demonstrated the vast superiority of the jet fighter. An Me 262 reconnaissance squadron commanded by Captain Braunegg, one of the best reconnaissance pilots on the Russian front, continued air photography for a long time, despite the Allied command of the air. The jet aircraft brilliantly proved itself in this branch also.

It is important to remember that we should certainly have succeeded in putting an end to the Allied bomber offensive if production of jet fighters had begun only six months earlier. At the beginning of February, 1945, Major-General Fred Anderson, Chief of General Spaatz's Operations Section, told General Eisenhower at his headquarters in Paris that unless the ground troops occupied Germany by June German production of jet and rocket fighters would make it impossible for the American air fleets of one thousand heavy bombers and eight hundred escort fighters to continue their raids by day.

But within ninety days the armies of Field-Marshal Montgomery and General Bradley had brought the war to an end.

It was too late to turn the fortunes of war in Germany's favour.

CHAPTER XVII

RETRIBUTION

GÖRING described 1942 at a secret conference as the nadir of the Luftwaffe. British air raids on the German homeland were mounting month by month. The German defences were less and less able to cope with them. German countermeasures against the British Isles were no longer effective owing to the lack of suitable aircraft and the numerical weakness of our air force remaining in the West. To all this were added the failures on all fronts and the noticeable diminution of the submarine campaign. On the 18th October, 1942, Goebbels comforted the nation with a great speech over the air in Munich:

"There are those who ask why we do not go in for reprisals. The answer is simple: it is because most of our bombers are needed more urgently at Stalingrad and in the Caucasus. Today it is more important to continue our offensive in the Caucasus, block the Volga and fight in Egypt than to oblige the English by dividing our forces. The winner in this war will be the side which knows how to concentrate its strength at the critical point in the battle. Our cities in the West and North-west have to stand up to hard blows, but I have come away from my many visits to them with the firm conviction that their morale is equal to the trial.

"Today we find ourselves in the fortunate position of being able to increase food rations. Our food position is slowly getting better. It is only a matter of time and patience for us to exploit the vast areas in the East and use them for our own purposes. It is simply a question of organization.

"The Have-Nots are already becoming the Haves and the Haves are becoming the Have-nots. We can be completely satisfied with the course of the war so far."

But it was not to turn out like that. The catastrophe at Stalingrad brought all those hopes to nought. During 1943 the enemy air raids did not ease off. Quite the contrary. Our war leaders had to stand by in idle fury while one German city after another was reduced to rubble. The nation began to cry out for reprisals. Göring had something to say about it:

"I could tell my fellow countrymen something which is hard to bear. I have gone in for reprisals already—and not so long ago.

At the moment the situation has certainly changed against us, and we are more knocked about than the fellows over the way.

"But even if all Germany's cities were levelled to the ground the German nation would live on. It would be pretty bad, but there was a German nation before there were any cities and men lived in caves. But it would be the end of everything if the Bolshevists got in. If Berlin disappeared from the face of the earth it would be terrible, but not fatal. The German nation has existed before without Berlin. But if the Russian gets to Berlin the German nation will cease to exist."

In view of the continuous, if controlled, withdrawal in the East this was a dismal look into the future, and not less dismal was the admission that the Luftwaffe was not in a position to carry out reprisals against England. How was the breach to be filled?

Propaganda got to work again. The "miracle weapons" were announced to all the world. They would not only be able to rub England off the map but, in Goebbels' version, would show themselves to be a one-hundred-per-cent guarantee of victory.

The propaganda itself was intended to work miracles. With its help a large section of the German nation was kept faithful to the cause in the last hard years of the war, despite all the defects and the desolation and suffering caused by the ever fiercer bombing war. When a retreat was announced or a defeat had to be admitted or an ally fell away or a popular hero lost his life, the "miracle weapons" changed despondency into defiance and doubt into confidence. After the Hamburg catastrophe and the loss of Sicily, for instance, Goebbels broadcast that on a recent visit to a war factory he had seen weapons which froze him to the marrow. And the Minister of Information and Propaganda was not lying this time, even though he had no idea when, and in what numbers, these novel weapons which were to revolutionize war would really be available.

Science and technology could not work miracles. Based as they are on the properties of materials, they must hold their hand until their discoveries and inventions, many of them quite revolutionary, have passed through the stage of experiment and development and shown that they can be applied for practical purposes. Patience and relentless toil are required to overcome the inevitable set-backs and disappointments.

But in the hour of supreme danger men cannot wait but seek desperately for a way out and hope for miracles. Hope was all that the German nation had left, and Goebbels fed it in masterly fashion. Giving evidence at the end of 1945, Colonel von Brauchitsch, Göring's aide-de-camp, paid him this compliment:

"On the air and in the press there was more and more talk of new secret weapons and things of that kind. No one could believe that all this was merely invented by propaganda, for it was unthinkable that in such a matter of life and death any government could deceive its own people, and on such a scale. It was only in that firm conviction that the civil population faced up to the ghastly sufferings of the bombing war."

What was the actual position with regard to the development of these weapons in Germany?

It is true that the German technologists had surpassed themselves in their efforts in this field. Revolutionary inventions, not paper projects, had materialized and proceeded to the production stage. I will confine myself to that group of these new weapons which is well-known—the V-bombs.

The V.1 was a flying bomb with a flat trajectory, powered by a ram-jet engine which was attached to it. Several concerns worked together to produce it. On the 5th June, 1942, it was offered by the Gerhard Fieseler Works to the Technical Department as a long-range weapon against London. Soon afterwards approval was given. With it land targets at a distance of about one hundred and ninety miles could be attacked.

Production of this new weapon was put in hand at once. The first tests took place as early as Christmas Eve in 1942 at the testing grounds at Peenemünde on the Baltic. There were sixty-eight firings in the early months of the following year and twenty-eight fulfilled the conditions required. In thirteen cases the pulse-jet did not function properly. A flight of 170 miles involved a consumption of 130 gallons of fuel. Both the range and the altitude were subsequently increased. Further tests suffered from the inability to supply sufficient carriages.

Our High Command pressed for immediate deliveries. Although the testing had not been completed, mass production began in the Volkswagen Works at Fallersleben in September, 1943. The risk that some of the bombs would prove ineffective was deliberately taken.

The tests between the 1st August and November, 1943, resulted in 150 urgent modifications—not less than 130 new parts had to be produced—but the test of the bomb as a whole in November gave such a result that serial production had to be stopped.

The technical difficulties were not the only ones which caused delays in the testing of the V.1. They were increased by the growing enemy raids on German factories. After one part of the production, especially of the prototypes, had been transferred to the Fieseler works at Kassel the latter was very seriously damaged by several heavy bombings. It had to be moved elsewhere. Fieseler under-estimated the difficulties to be encountered in the new works. Hitler had issued an order that no foreigners must be employed in the manufacture of the bomb and so Gerhard Fieseler had to dismiss forty-five per cent of his labour force. It was weeks before sixty per cent of his workmen were actually working at the new site and it was months before the Fieseler concern could produce more than two prototype bomb bodies a month. They were then assembled at Peenemünde by special labour units from the Volkswagen works. A heavy raid on the 17th August completely paralysed the testing ground for five weeks.

The indirect damage caused by air raids on the accessories factories was as serious as the direct. In the autumn of 1943 only thirty per cent of the directional and navigational equipment was being delivered. The lack of skilled labour was making itself felt. In the Volkswagen works the labour force was 1,453, of which 1,059 were Germans. It was nothing like enough. To make matters worse, at that time air armament was being grossly neglected both by the chiefs of the Wehrmacht and the Ministry of Armaments and War Production. When manpower shortage and difficulties of delivery were being discussed at a conference on the 30th November, 1943, General Vorwald, Milch's representative in the Technical Department, remarked: "Saur really makes it his job only when he is in complete charge of anything. If it was he who had to produce the V.1 and the Me 262 he would be in it one hundred per cent. But as long as it is we who have the responsibility and he has only to deliver the goods he does not go all out." So in October we were short of more than 250 skilled men and 600 to 800 machines, which were not forthcoming until the office furniture and typewriter industries were closed down.

The difficulties of conveying the bombs—which were twenty-nine feet long overall—were also very great at that time. Railway transport was slow owing to bombing and this again delayed the testing. Sufficient motor transport did not exist.

Göring again clamoured for the immediate employment of the V.1, especially as the V.2, which the army was bringing out, was to be ready by the end of the year. At a demonstration of the newest weapons at Insterburg airfield in November, 1943, Hitler was shown the V.1. To his question when it would be ready for use the reply was that it would be March, 1944. General Bodenschatz, the Luftwaffe representatives at Führer H.Q., commented subsequently: "Who was the pessimist who arranged this meeting?" As a matter of fact the date had been fixed in ignorance of the fact that serial production had been stopped as a result of poor results on test.

The position was thus in fact worse. The plans of the Technical Department provided that 1,400 V.1s should be produced in January, 1944, and 2,000 in February. In fact nothing was available.

Meanwhile the High Command of the Wehrmacht had ordered a conference for the 3rd–4th January. It was attended by General von Axthelm and Staff Engineer Bree who reported progress to Colonel-General Jodl, Chief of Staff, and General Warlimont, head of the Intelligence Service. The outcome was that Hitler fixed 15th February for the début of the V.1.

Serial production was resumed and it permitted about 300 V.1s for March, 1,000 in April and then 1,500 to 2,000 per month onwards. At any rate that was the intention. The tests of the firing points had shown that the ramps must be lengthened. The 200 ramps considered the bare minimum were not ready before May though they had been given the highest priority.

In northern France, by the autumn of 1943, the ground organization comprised eight major installations and ninety-six field launching sites. These concrete firing points had to be well camouflaged so that they could not be discovered prematurely and destroyed from the air. This presented great difficulties, which were never overcome.

Before the constructional work began, Field-Marshal von Rundstedt, the commander in France, pointed out that there was insufficient constructional material as the Atlantic Wall had absorbed it all. Thirty-five thousand men, mainly French, were occupied in building the sites.

Up to December, 1943, the work was hardly affected by air attack but from then onwards they were under continuous fire. Of the labour force 20,000 men absented themselves. While German soldiers were forbidden to go near the sites for security reasons, the employment of Frenchmen left the field wide open to espionage. At a certain stage of construction almost all the sites, however well camouflaged, were attacked from the air. At the end of December fifty-seven per cent of all the sites had been under fire; ten per cent totally destroyed; ten per cent seriously and ten per cent moderately damaged. This development drove the High Command to a simpler type of installation, as complete repair took six to eight weeks. Ultimately the Flak General's idea that widespread dispersion of the sites was the only protection against mass attack from the air had necessarily to be adopted. It had the additional advantage that it avoided the colossal expenditure of material on concreted sites.

As movement in France was impossible except at night the work was subject to long stoppages and delays. To camouflage the actual firing points we mounted on lorries apparatus which could imitate

12

the flash and sound. The sites were served by men from the flak artillery, who had already assisted in the production and testing. In August, 1944, these flak formations were formed into the 5th Flak Division, which was eventually taken over as the 155W Flak Regiment by the S.S. General Kammler.

The V.1 was first employed against London on the 13th June, 1944, four months later than Hitler had decreed, and a week after the invasion of France had begun.

Over a thousand V.1s were launched within the next eight days, of which:

Twenty-nine per cent reached the target;

Forty-six per cent were brought down by the British defences (twenty-four per cent fighters, seventeen per cent anti-aircraft guns and five per cent by balloon barrages);

The remaining twenty-five per cent got off course and went astray, most of them probably ending up in the sea.

The figures speak eloquently for the efficiency of the British defence against the V.1. As the German leaders selected the London area for this "retribution" measure it was easy for the British to establish very quickly a defence zone which by the beginning of July was manned by thirty-six A.A. regiments, with 1,000 guns and 500 barrage balloons, and a big fighter contingent. By September, this zone comprised 2,800 guns, 2,015 balloons and a much larger concentration of fighters, so that by the end of the offensive only nine per cent of all the V.1s were getting through to the target area. The high speed of the new British Tempest fighters enabled them to tackle a bomb flying in a straight line at about 400 m.p.h.

The loss of our northern French bases gave London a temporary respite from the V-bombs.

The development of the V.1 was originally based on the idea of relieving our bombers, economizing in men and replacing expensive aircraft by cheap flying bombs. Air attack on the launching sites and the imminent danger of invasion inspired experiments to launch the V.1 from aircraft. At an altitude of 650 feet it could be fired from an He 111 bomber and theoretically reach its flying altitude of 1,150 feet under its own power. But this technique was strongly opposed by the Reichsmarschall's technical adviser, Colonel Diesing. In his opinion "the Luftwaffe was not in a position to use bombers for that sort of thing".

The launching of the V.1 from aircraft began seriously only when the German ground organization in France was lost as the result of air attack and the progress of the Allied invasion. In the autumn of 1944 our He 111s used to take off from the Oldenburg area, fly to a fixed launching point over a Dutch island and then launch their missiles. But the English night-fighter defence caused such a heavy loss of aircraft that systematic attack had to be called off.

Of 1,200 V.1s launched from aircraft against London only 205 reached their target. Better success was obtained against Antwerp between the 12th October, 1944, and 30th March, 1945, when a daily average of thirty-four bombs did a great deal of damage in the Allies' principal port. During the Ardennes offensive the V.1 was also used against targets in the back areas of the American First Army.

The experiences of the war showed that even in that stage of its development the V.1 was a military weapon which could be used with success against ground targets such as ports, cities, troop concentrations, industrial centres. For purely technical reasons such as limited range and accuracy and lack of sufficient speed, hopes placed on it by the German leaders and the effects as a strategic weapon of vengeance promised by our propaganda could not be realized. Not before the further development of jet propulsion did the ideas embodied in it mature.

Although the V.2 (officially A.4) was a purely army development, and was handled by the army artillery throughout its short career, it comes into the group of those new weapons which will make a future war a space war.

The first experiments in Germany with liquid rockets of the V.2 type date back to 1933. Limited resources and technical difficulties kept their prototypes A.1 to A.3 in the preliminary stages in the pre-war period.

In contrast to the V.1, the V.2 was developed as a rocket missile guided by an electric beam. It had no wings like the flying bomb but only stabilizers and fins. The propulsion of this giant missile, weighing over twelve tons in all, came from a combustion chamber fed with liquid oxygen and ethyl alcohol. The burning gas masses streamed out through a single exhaust in the stern and created the mighty thrust which shot the great body, with its tanks and auxiliary machinery, propulsion unit guidance apparatus and 2,150 lb. of explosive straight up into the air. At a height of eighteen and a half miles the missile attained a speed of 5,200 to 5,500 feet per second and was still rising at an angle of forty-five degrees. The combustion period in the chamber was sixty-two to sixty-eight seconds and a maximum altitude of nearly sixty-two miles was achieved. With a total flying-time of 320 seconds the range was between 170 and 210 miles. The accuracy of these pre-set guided missiles was such that fifty per cent should fall in a circle with a radius of fourteen miles.

Between the 8th October, 1944, and the 2nd April, 1945, we fired 1,115 V.2s of the A.4 type against England and 2,050 against Brussels, Antwerp and Liège. In this type steering control was limited to initial direction and control during the combustion period.

The enemy's only defence against the V.2 was systematic bombing of the launching sites and the factories where they were produced. The great speed of descent—2,000 miles per hour—precluded fighter or A.A. defence in the vicinity of the target.

In the years after the war the V.1 and V.2 have undergone further development in the U.S.A. and Russia. Most of the German V.2 scientists are working at the rocket testing ground at White Sands in Texas. After the war President Truman said that the value of these German inventions is not to be estimated in dollars. Those scientists were snatched from the clutches of the Russians at the last moment.

The Russians too are working with might and main on jet propulsion, rocket bombs and the technique of long-range direction. The fact that they have transported a whole staff of German scientists and experts into the interior is further proof of the importance of these technical and industrial novelties.

CHAPTER XVIII

IN BERLIN

"WE met this morning as arranged. It was our first real meeting as on previous occasions we had only exchanged a few words on matters of no importance.

Just before 5 a.m. I am at the dark and desolate door of No. 53, Pariser Plaz. A sleepy old porter lets me in. A black B.M.W. car is standing in the courtyard with its engine warming up while the driver cleans the windows. No one takes any notice of me, though I appear to be expected."

A mysterious telephone call on the previous day had summoned me at this unusual time to an even more unusual meeting with the man whom Hitler called the most important man in Germany's war effort. Albert Speer, the Minister for Armaments and War Production and Deputy Transport Minister, wanted to have a talk with me. I had no inkling of his reasons. What could he want with me?

A door banged to and someone came striding towards me through the inky darkness. I instinctively left my corner and stepped into the meagre light from the car's lamps. "Good morning. Now let's go," said a voice in the dark. Then we were facing each other. Speer's tall broad figure, with the unusually big head. He was wearing a grey civilian overcoat but neither hat nor gloves. "I'm driving to the Western industrial area. Will you come a bit of the way with me? I'd like to talk with you." Brief instructions to his chauffeur, who was to follow behind with my car, and then the car door snapped to. We were alone.

Speer drove himself. The headlights blazed, taking in the façade of the Adlon and the narrow passage through the Brandenburger Tor, and then we passed the Siegessaule, put on speed and raced down the east–west axis.

Rain and snow beat against the glass, to the accompaniment of the rhythmical swish of the screenwiper. The enormous black lamp standards of Adolf Hitler's wide parade ground stretched their bare dead fingers to the most depressing sky Berlin ever gazed into. Here was the scene of Mussolini's brilliant reception and the great military parades on Hitler's birthday, the 20th April; here, the *via triumphalis* of the Third Reich, the low-flying Luftwaffe squadrons had swept past their Commander-in-Chief.

All that is of the past. Dead, finished, played out, lost. The air terror is paralysing this city, squeezing out its last breath. But the lamp standards are still here. They fly past us, at meticulously regular intervals. Only a few are a bit battered and bent. They are not lit these days. Their eyes are closed. Torchlight and flood-lighting are things of the past. In the darkness these lamps seem like gallows, an endless row of gallows which have grown out of the ground to cast giant shadows over the whole country.

Neither of us spoke.

I could not think of anything to say and it gave me a weird feeling sitting next to the big silent man driving his heavy car over slippery roads as if it had been a toy.

"Don't you agree that the war is lost unless something very drastic is done?"

I started as if I had been clubbed with the words. Had I been dreaming or had I really heard him say: "Don't you agree that the war is lost?"

He looked hard at me, waiting. Then the voice was heard again, quiet and clear but very far away: "Don't you think the war is being lost?"

I bit my lips and forced my thoughts into some sort of order, but Speer resumed before I could reply: "I know your history, and that you've always had to tell your superiors and Göring the truth. You can be quite frank with me. That's why I've asked you to come along. We can't be overheard. Anyhow, we are colleagues in the same line of business"—here he smiled—"as I've read some-where that you want to be an architect."

The blood rushed to my head. I made up my mind that I would keep nothing back from the man who I knew had been one of Hitler's closest associates for nearly ten years before his appointment as Minister of Armaments. I would tell him of our distress and sacrifices at the front.

I stammered at first before I got going. I told him that we were back to the most desperate days of the air war, with hundreds and hundreds of our comrades missing or falling or being shot down. I spoke of our sense of being abandoned between earth and sky, the loneliness of the North Sea, the deadly chase in the Mediterranean and Atlantic, the scandalous tactical methods imposed on us in recent years, the distress and cry for help of this young generation which fanatically believed it was fighting in a just cause. It had entered the war thoughtfully and soberly, not with reckless jubila-tion. Our young men felt subconsciously that great changes were called for. Their faith had been shattered in the last few years, as had their trust and confidence in Göring and the Luftwaffe leaders.

As always the best had fallen and were being driven to their deaths by a small caste sitting safely at their desks.

I also spoke of the vain attempts since 1941 to bring about a change by conferences, reports and proposals, not forgetting my last letter to Hitler; and I did not leave out the intrigues among Göring's entourage, the background to the suicide of Udet and Jeschonnek, Göring's weakness in dealing with Hitler, Goebbels' propaganda about secret weapons—which were not ready.

I got more and more excited and my accusations became more and more vehement. Visions of these years of my youth, the accusing looks of my dead comrades, passed before my eyes and seemed to suffocate me. The sweat poured from my brow. This endless chain of death. This damned war without pity.

"I was expecting you to be completely frank. There's no one I trust more. I shall be back in Berlin in two or three days. We'll meet again at once."

The car drew up under a bridge over the autobahn and only then did I become aware that we had been driving for over two hours and I had been talking practically all the time. It was raining less heavily and gradually becoming light. Until my car caught up with us we had the hood down, despite the cold and wet. It was a relief. Clear and cool heads are wanted now and in the immediate future. I should like to have known the unspoken thought behind that great forehead with the bushy eyebrows.

"We met at Peenemünde last summer, at the V-weapon demonstration. I hope we shall see much more of each other. Good-bye!"

He drove off, going west, while I returned to Berlin.

I was still more agitated than I could have thought possible. Was what I had said true? What would the immediate future bring? Then the air-raid warning brought me back to stark reality: "large enemy bomber formations heading for Berlin. Leading aircraft near Brandenburg. . . ." The ghostly terror, war's fury, was once more suspended over Berlin. How could we put a speedy end to it?

Speer sent for me every day, usually at night. A last attempt was to be made to destroy the Russian power plants by a comprehensive strategic air offensive. Since taking office as Armaments Minister, Speer had stood for the idea of continuous and persistent economic war against Russia, but had not succeeded in forcing his plans through in the absence of that backing from Göring which was then still necessary. Now Speer took me straight to Hitler. I was to tell him all about the new "Eisenhammer" plan.

A great part of the Chancellery building had been destroyed by bombs. We went in at a side entrance and Speer took me by a roundabout way to the Führer bunker. No one besides Colonel von Below, the Luftwaffe A.D.C., was to know that Hitler had sent for me.

We had our conference in a little room in which the space was almost taken up by a square map table and a number of landscape pictures propped up by the walls.

I was completely taken aback by the physical changes in Hitler. He was years older, and stooped like an old man, and his arms and hands trembled so much that at the end of the meeting Speer got his signature in the form of a rubber stamp.

Two days later I was back in the Führer shelter this time with Göring as well as Speer. From then on I shuttled back and forth between the Chancellery building, Speer, Himmler, Karinhall and Goebbels. Yet I was still Commander of what was left of the bomber force. Long-range aircraft, special weapons and, to crown all the "Suicide Group" were brought within my province.

When I turn the leaves of my diary today, those grotesque figures from desperate days pass before me. It certainly was an act of desperation when certain circles in Germany recommended the imitation of the Japanese *kamikaze* flyers. So much has been written about these German *selbstopfermänner* ("Self-sacrifice men") that, as their commander at one time, I will let a few extracts from my own notes speak for themselves:

"Heiner and Hanna (Reitsch) were there. This latest idea was aired again. The Suicide Group were this time represented by Captain Heinrich L———, familiarly known as 'our Heiner'.

"The Suicide Group are volunteers who are prepared to throw away their lives to save the Third Reich from collapse. These must be the best of the best, the most fanatical of the fanatics, the bravest of the brave, the fieriest of the believers, and their leaders will stare at you with tears in their eyes and say: 'we stand under a *higher law*!' That, at least, is what one of them said to me when he was brought before me for overstaying his leave. But for the time being they were under my command.

"How did they come to be assigned to me? In the summer of 1944 Knemeyer told me a tale of woe. He knew the story of the Suicide Group men. They were idealists, with decent and useful men among them, who believed what they were told. Some of them had been more or less pressed into this formation by Skorzeny. When I spoke to them later they did not really know to what they were committed. They thought they were to take part in some 'special operation' like the rescue of Mussolini and would get a medal. But they wanted to live, not to die! *That* I could well understand. There is an old saying among airmen that the closer a trainee clings to life the more likely he is to survive. It is particularly true of war. So there were no ex-bomber and fighter pilots among these men.

"The idea of hurling oneself at a target with total disregard of consequences is not new, however foreign to our mentality. In the age of modern scientific warfare it is not even the best solution, to put it mildly. There are more effective methods. But the circulation of such an idea, and its results, are dangerous.

"About this time Milch had agreed that the V.1 bomb should be modified so as to be manned by a live pilot: Knemeyer, who had a wonderful reputation as airman, technician and ordinary human being, had previously urged the folly of such a modification on Milch; it would mean certain death to its pilot. Milch was convinced by his technical arguments and said that such an instrument of 'total war' should not be built. Two days later, Otto Skorzeny went to see the Field-Marshal, and on the same day the latter issued an order that the modification of the V.1 should be speeded up.

"Such was the situation when Knemeyer said to me: 'You must take over the whole outfit. You're the only man who can stop this nonsense, because you can talk to Hitler and the S.S.'"

About two months later:

"Herr Baumbach, what's the position with the Suicide Group people? Three months ago I once had them here in the Ministry. I was told they would be employed in the next few days. I actually gave them parting presents. Since then I've heard nothing. I've been waiting in vain for good news. What is the position with these men?" Dr. Joseph Goebbels, the Minister of Information and Propaganda, waited for my answer. He was sitting next to me at a round table in his room, wearing a smart, double-breasted civilian suit and a most fashionable tie, and was holding a slim silver pencil and a silver-edged notebook.

A troublesome question. Goebbels lowered his mask a bit. He said he had been told that these men were to carry out an operation which would decide the war. Accordingly, he had made a speech about the meaning of sacrifice for the Thousand Year Reich and said their names would be remembered for ever. . . . "You know the sort of thing." I was simply astounded at the cold tones in which he spoke of "the great truths". What he meant was that he had not believed all he had been told and tried to give them a shot in the arm. I thought I could detect a fundamental contempt for mankind in his words. Then he turned to something else and an hour later I left the Ministry.

I appointed Gottlieb Kusche as commander of the Suicide Group. The war was approaching its end with giant strides, and certain circles which had no use for the Luftwaffe were even more insistent that it was time to use the S.G. The hangars at Prenzlau airfield were stacked with V.1's which had been insufficiently tested. Heiner was not to take part in the operation because it was his job to train the Group recruits in "the great truths". And Hanna Reitsch, once an expert glider pilot, had to pilot Colonel-General Ritter von Greim's aircraft on his tour of inspection.

Meanwhile two rivals had appeared on the scene. Colonel Herrmann, who had once produced the "Wild Boar" (searchlight barrage), organized a new ram commando at Stendal. The pilots were to ram the four-engined bombers with their fighters. Major-General Storp, the last Inspector of Bombers, wanted to start a great recruiting drive and form a Luftwaffe Suicide Group division. The men were to wear army uniforms—to throw the many enemy spies off the scent—and the division was to be called the "Hermann Göring Fighter Division". Could idiocy go further?

After an almighty row between Storp, Major-General Pelz, Colonel Knemeyer and me in Göring's presence, I spoke to Speer next morning. He alone could get us out of this maddening predicament. He took the bull by the horns and hauled me off to the Chancellery.

The Führer rejected Suicide Group operations. He had already told Hanna Reitsch that he would have none of it. The German soldier must have a chance, however small, to come through.

Major Kuschke saw me that night and the very next day the S.G. were transferred, some to their units, others to flying schools, training depots and so forth. The next few days were to show how necessary such precipitation was. I was told that I was to put S.G. operations in hand. The Führer must not be "bothered" with such things. The Reichsführer would shoulder the responsibility. S.G. at any cost!

Then I had a talk with Himmler alone at his field headquarters near Prenzlau. Ultimately he accepted my arguments.

Meanwhile, Gottlieb Kusche had done fine work in the Berlin-Alexanderplatz. The suicide pilots were scattered to the four winds, where no order, no Heiner and Hanna could get them back. Most of them were only too glad and I myself was delighted to be rid of a nightmare.

The reader may smile at the above or consider it incredible. Unfortunately, it was the bitter truth. My refusal to collaborate in such idiocy—and even command it—was within an inch of costing me and my officers our heads. Only those who lived in the hyperneurotic atmosphere of Berlin in the last months before the collapse are in a position to understand and judge what happened. Under the terror of the daily raids, the effects of endless defeats in the field and the collapse of the home front, many men ceased to think and act normally. It does not mean that they should now be judged by cheap, normal standards. After further remonstrances by front-line commanders continued to be without effect on Göring, and even Speer at the end of 1944 was unable to move him to take the measures which the German nation vitally needed, I found myself driven to make a supreme effort with the Reichsmarschall. While we were travelling together from Karinhall to Berlin in his car, I handed him the following memorandum:

Operation Headquarters, the 9th January, 1945.
Only for Herr Reichsmarschall personally.
Herr Reichsmarschall,

I am still reflecting on my talk with the Herr Reichsminister Speer yesterday morning. It is the final reason for this appeal which, in this precise form, has been on my mind for weeks.

Since my first memorandum in August, 1941 I have repeatedly set forth my fundamental ideas about this war and its course. Unfortunately, without success. My only "success" was that from time to time I was thrust aside as a bore and a nuisance, notably after the death of Colonel-General Jeschonnek, for which the Air Staff and the close associates of the Herr Reichsmarschall were responsible. That is a fact which is an open secret in the Luftwaffe.

After careful consideration I have felt that I must present this memorandum. I shall say nothing about my active service. But I am determined to be entirely frank, without regard to any consequences to myself.

In my memorandum to the Führer a year ago I said I had never heard of any general or higher staff officer leading his formations to battle, or of any Luftwaffe general losing his life in action. At that time I gave you my views quite openly at a private meeting with you in your study at Karinhall. I was anything but silent about the shortcomings of the Luftwaffe, and particularly the higher leadership and the air armament industry. I also said that after the Anglo-American invasion of North Africa and Italy the war could not be won in the field.

At first I was just shouted at: "Well I'm damned! Pessimist! Where's your crown of thorns? Get out!" But a few hours later you told me that you knew all about those shortcomings and would not only remedy them but yourself look more closely into the affairs of the Luftwaffe. The famous Areopagus in Gatow, which was your own idea, finally destroyed all hopes. That was 1944.

What has happened since is too well-known. I will only confine myself to a few observations.

After the events in the Mediterranean in the winter of 1943–44 I was transferred to Berlin. The new Inspector of Bombers, who was one of my best active service comrades, was, on the Führer's orders, appointed as "Commander of the Offensive against England", to conduct the "retribution" raids on London and the British Isles. It was in vain that I and other prominent airmen such as Heilbig tried to dissuade him, even promising to back him with our own heads. He himself was doubtful about the success of his mission.

So in 1943 the bomber force was betrayed and its great hopes of improvement and reorganization grossly mocked. As the flower of the bomber officer corps had already fallen in action the losses of the crews employed against England and the Mediterranean mounted steadily, though Major-General Pelz's leadership seemed to be better than that of the old generals. The material and technical superiority—not to mention superior numbers— of the Anglo-Americans was in any event bound to frustrate any rational and effective air strategy. But decisions which were not compromises were never taken. "Improvise!" is now the order of the day, and now we have improvised to death because we have refused to realize that we have long been fighting a poor man's war!

At a conference called by the Quartermaster-General in August, 1943 at which I was present as General of the Bombers, I said to Field-Marshal Milch (I quote from the transcript): "As we must first make our home defence strong, I propose that in view of the circumstances we drop production of the He 177 Bomber and use the manufacturing capacity thereby released for the rocket fighters, so that home defence in all its aspects shall be as strong as possible." Field-Marshal Milch evaded a reply and passed on to something else. When I continued to urge my views, on guided bombs for instance, no more attention was paid.

After my own experiences in Norway, the Atlantic and the Black Sea, Sicily and against England, I was positive that my homeland would very soon be called upon to endure the heaviest enemy bombing, so that it seemed to me that the first duty of our leaders, and particularly responsible Air Staff, was to create a fighter force for the defence of Germany.

Unfortunately, I and hard-thinking officers such as Knemeyer and Galland, have proved right. But we got no hearing or were without the necessary backing. Today it is too late. Towns and villages are being reduced to dust and ashes and every day thousands of people have to die a terrible death. When we, front-line fighters, proposed in 1942 that the bomber force should be reduced for the benefit of the defensive air force we were told off roundly by the Commander-in-Chief himself. I still possess the written record in which occur the words: "I will not hear of any reduction of the bomber force! I shall double it!"

Since my first meeting with you and Colonel-General Jeschon-nek in the summer of 1941 I have never been backward with my opinions, and ended up at the Areopagus at Gatow with an attack with open sights on "the generals" as a body and this so-called general staff in its red trousers. By that I meant the immediate entourage of the Reichsmarschall. I was the man who officially demanded that it should disappear.

Herr Reichsmarschall!

Today only two things matter to me:

(1) My personal duty to my country which, at the beginning of the war and when I was full of idealism, I thought I could best serve as an airman;

(2) The honour of my little family which I would defend with my last breath.

I recognize no other duties or obligations. It is no longer possible for me to allow my name to be nothing more than a sign-board. I have therefore not complied with an order I have received from the Luftwaffe Personnel Office to attend and speak as a holder of the Knight's Cross at a political demonstration at Berchtesgaden. I fully realize the possible consequences, as I have received several warnings from Albrecht, at the Party headquarters. I consider myself solely to be the mouthpiece of the bomber crews. Most of them are no longer with us, but those that are look up to me to represent their views without shrinking. I am doing so. I stand by everything I have said. If it is too drastic, please put it down to my youth.

It is possible that my views cannot be accepted and that I must be sacrificed for so-called "reasons of state". Even so, immediate action is urgently called for. Not an hour must be wasted if it could help to save the Fatherland. That is now the sole task of the country's leaders!

I should be an infernal coward if I could not screw up the little moral courage required to do with my voice what thousands of my comrades have done with their blood.

As I do not wish to be running around as a marionette, I hereby place my rank and the orders I have won in battle at your disposal.

(Signed) Baumbach.

I might have saved myself the trouble. Nothing whatever happened. Göring continually postponed any discussion of my memorandum. He had other personal worries.

So I had to act on my own. In those last weeks in Berlin, Speer was the life and soul of all the forces which felt themselves responsible for Germany. Like the Knight without Fear or Reproach he was always striving with all his might to prevent Hitler from destroying Germany. In those last moments he averted untold damage and disaster. Of the top leaders he alone, supported by a few officers and industrialists, saved us from complete chaos. Yet his friendship with Hitler was particularly close and how difficult it was for him to decide and how much it cost him later on to open his defence with the words: "If Hitler ever had a friend it was myself. . . ."

On the 15th March, 1945 he produced a memorandum on the economic position which ended:

"In this stage of the war we have no right to carry out demolition work which will injuriously affect the life of the nation. If our enemies mean to destroy this nation, which has fought with unexampled courage, that historic act of infamy must be theirs alone. We have the duty to leave the nation all the possibilities of recovery in the distant future."

In a private conversation with Speer Hitler told him that he could not accept his reasoning. If Germany were to lose the war, the future of Europe would be in the hands of the Eastern nations. In that case, the German people deserved no further consideration; the best had already fallen and only the less worthy were left. On the 19th March, 1945 he issued his "Demolition Order".

"All military, transport, postal, industrial installations, food supplies and valuable property within the Reich which the enemy can use, either immediately or in the near future, to carry on the war are to be destroyed."

This demolition order would have involved the complete collapse of Germany's industry, economy, transport and national life.

Extract from Speer's memorandum to Hitler of the 29th March, 1945:

"I can only carry on with decency, conviction and belief in the future if you, my Führer, dedicate yourself to preserving the vital resources of the nation. I am not referring particularly to your demolition order of the 19th March, 1945 with its precipitate measures, which must take away our last industrial possibilities and cause the greatest consternation when they become known. These things are, of course, vital, but lie outside the fundamentals."

It was too late. The twilight of the Gods had already descended on Germany.

In those eventful days we had asked the Berlin Philharmonic. which everyone had almost forgotten, to give us some concerts, They played for the Berlin volunteers once a week, and in the evenings Wilhelm Kempff and the young violinist, Gerhard Taschner, played for hours in our little circle of friends. Wilhelm Fürtwangler, who also wanted to come to Berlin, was on a black list with other eminent persons who were "doubtful" and were to be gaoled. At the last moment we had got wind of this and were doing everything to keep this great European out of the clutches of death. He was able to seek safety in Switzerland.

Then came the last concert. No one knew what was to become of this famous orchestra. Some notes in my diary bear witness to the heavy atmosphere of those days:

"The last concert.

"It was the days before the Russians came to Berlin. I had driven into the city once more to hear the Berlin Philharmonic play for us for the last time in the Beethovensaal. The programme— Beethoven's violin concerto and Bruckner's fourth symphony.

"Berlin, our Berlin, for the last time. As I walk through the streets in my shabby greatcoat, without badges of rank, I suddenly realize that I have never before seen anything so grim, so poverty-stricken, so utterly martyred and forlorn. There was another air raid this morning and the city is burning from end to end.

"Low, heavy clouds lie over the silent trees in the Tiergarten and eventually it begins to snow. I pause and listen to the whisper of the flakes to last autumn's leaves. The titmice squeak as they scurry across the paths, but otherwise there is no sign of life on this benumbed, tortured little spot.

"I walk past the Brandenburger Tor to the Potzdamer Platz. My eyes are full of smoke and dust. The women's noses and mouths are swathed in scarves and the men huddle in their coat collars and pull their hats down over their eyes. Berlin still lives, though life means no more than just existing.

"Half an hour before I had said good-bye in the Lietzenburger Strasse to a young designer. I was terribly shocked at her appearance and even more by the change that had come over her in just a few weeks. She said she meant to stay in Berlin—even if the Russians came. Where else could she go? No, if the Russians came, it would at least mean the end. There was a lot to be said for it. The first thing to happen would be the abandonment of the butter shop opposite. They would get enough to eat for once.

"What war can do to people!

"I enter the Beethovensaal. The Philharmonic's last concert. No charge for admission today. Anyone can go in, and the Berliners troop to their Philharmonic. In the unheated hall the audience keep their hats, coats and gloves on. Many have to stand. Working Berlin is here for this concert. The Berliner who cannot live without his Berlin. My heart thumps and then I slump into a seat and wait.

"Beethoven. His music enters my soul and gives my feelings wings. The hard and the soft, strength and weakness, the demand and the request, love and brutality, rejoicing and tears, the diabolic and the godlike—all jostling together. I see trees bending in the wind, flowers breathing in the sun, waves breaking on the shore and clouds drifting far, far away.

"My eyes wander over the faces of the audience. I see many who look spellbound, dedicated, and I suddenly realize that I and all these people whom I do not know are caught up together in the same spiritual experience. And then the concert room seems to be the vault of heaven and I see an unending procession of men become spirits streaming in the same sense of comradeship, but far wiser and happier, towards the perfection of divine unity.

"And then Bruckner—eternal marriage bells. With this composer every work is devotion, goodness, joy, providence—and each a rare emotion. Love, a bridge between one man and another. It is humanity's greatest gift.

"The concert ends. No one in the hall is ashamed of his tears. There is a deathly silence, the highest reward that a transfigured audience can bestow. It is all over and we must go. It is as if all these people do not want to separate or break up the community of spirit."

Then came the final hour for the capital, bleeding from a thousand wounds. The Red Army's iron ring closed round "Fortress Berlin", which was in fact no more than a giant bomb crater. There was nothing more to save. Berlin was left to the Russians. But senseless destruction in West Germany could still be averted.

When Speer on the 23rd April flew into Berlin in a Fieseler Storch aircraft and spoke to Hitler for the last time, the city was completely surrounded by the Red Army. Hitler said he would have much preferred that it had been the English and Americans rather than the Russians. He withdrew his demolition order but could not be moved from his decision to die in the Chancellery, still hoping that Berlin would be relieved. "A comet burnt up in its own flames."

CHAPTER XIX

BEFORE AND AFTER THE END

"Dear Winnetou or Kne[1]

"It is 4 o'clock on Saturday, the 21st April and we are south of Scheerin on the road to Lübesse and waiting until 8 o'clock. If anything happens, Major Strack, Lübeck, Plönnerstr. 8. Tel. 22332 will know. Ask for Colonel Holzhäuer.

"Food in any quantity desired can be obtained from my deputy, Wolf, in Hamburg (ask at the Governor's residence or the Armament Inspection Office for his address) as he is in charge of supplies to Hamburg. Instructions to Wolf enclosed. In a hurry owing to the situation at the front. If meeting not possible owing to situation, we meet in the forest between Seegeberg and Bad Bramstedt at the same time.

Yours,
Speer."

In this note Speer arranged our next meeting after Berlin had got too hot. We had fixed up long-range aircraft and flying-boats so that we could get anywhere on earth. No one yet knew exactly how the war would end. The Allies were broadcasting all sorts of promises to the German people if they would lay down their arms but their low-flying aircraft were mercilessly strafing the fleeing refugees on the roads and bombing towns and villages wherever their troops met with the slightest resistance.

In the last days of April we were in Hamburg.

"It is over. Hamburg will surrender in the Town Hall. Yesterday evening, modern Winnetous and Shatterhands, we slunk into the city with our speedy B.M.W. cars, machine-pistols and God knows how many passes for the street barriers and traffic controls.

"The residence buildings lie between gardens and a high wall in the darkness of a side street near the Alster. 'Halt!' We switch on the lights. In the headlights I can see the safety catches of machine-pistols being gingerly withdrawn and look into the uncertain young faces of the eighteen- or nineteen-year-old guards.

"Our cars come to a stop with a jerk. I fling open the door and call out my name. The sentry whispers something to his colleague. 'Pass!' I jump out of the car and give the young fellow a

[1] Kne—Colonel Knemeyer.

playful slap on the shoulder. 'No panic. They know all about us. You needn't wet your pants,' I add in *patois*. The sentries have recognized me and laugh.

"Carbines and machine-pistols are ten a penny in the corridors and on the stairs. In the obscurity of the long corridor the atmosphere is positively warlike. It is probably about half an hour after midnight. We are taken straight in. A thick-set little man whom I see for the first time gets up to meet us; it is Karl Kauffman, Gauleiter and Governor of Hamburg. He looks terribly harassed and short of sleep. We seem to have got him out of the folding bed in his office. My companion introduces me. We sit down. We get a brief résumé of the situation:

" 'The English are at the gates of Hamburg-Harburg and could fire on us at any moment. Resistance would be a crime. The immediate answer would be the heaviest bombing and many more thousands of helpless and defenceless people, particularly women, children and old folk, would lose their lives. Without regard to my own position I have only one duty left—to call on the civil population for peace and order so that the city, harbour and Elbe bridges can escape damage. That's what matters most at the moment— to save the Free Hansa City of Hamburg from complete destruction.'

"Kauffmann's expression, and the few questions he puts to us, betray the inner conflict between his innate common sense and better feelings and the Demolition Order. After listening quietly to our eye-witness account of the last weeks and days in Berlin his mind is made up. The man has triumphed over the party official. He tells us that his first responsibility is the fate of the civil population of Hamburg:

" 'I will surrender Hamburg at the Town Hall.'

"The next few hours show that 'Kuddel', as Kauffmann is called down at the docks, will keep his word. While we get down a bowl of soup Speer gives him a prepared broadcast to read. Half an hour later he and Speer are in a car driving through blacked-out Hamburg to the radio-station.

"The supervisor and engineers open their eyes wide when they hear that their wax discs are bespoken twice in the middle of the night but they are too sleepy to realize the full meaning of what is happening. They are told that in any case absolute silence is best. They have understood. By dawn all is over.

"I have a talk with my father's friend, Geheimrat Bücher of the A.E.G., who has escaped to Hamburg. My weariness has vanished. We pick our way out of the city through Uhlenhorst. My local knowledge comes in useful and we arrive intact at our caravan on the Plöner See. After this successful enterprise I am ready for the bottle of champagne which Harald Lechenperg gave me on our dramatic departure from Berlin. We drink to the health of the ancient free Hanseatic city of Hamburg. May its ships and its trade soon recover their lost prosperity. That was our devout wish this morning."

At Travemünde airfield the same day I was given a message that Himmler urgently wanted to speak to me.

My diary, 28–29 April, 1945:

"Thank God that's over! I would rather leave some things unsaid but it occurs to me that these diary notes may one day shed a little light on the strains, the desperate situation and maddening hurry of the last few days. So I will give some account of thi episode, my last talk with the most feared man in Europe, Heinrich Himmler, Reichsführer S.S., Head of the Gestapo and Chief of Police, Minister for Home Affairs and Commander-in-Chief of the S.S. Army.

"Hardly a week has passed since I was sitting with Himmler long after midnight, following an afternoon in which orders for my arrest had been given in Berlin.

"On the 20th April Himmler had not gone to the Chancellery for the birthday celebrations as usual. On Göring's instructions I had had private talks with S.S. Brigadeführer Schellenberg, the Chief of Military Intelligence. Schellenberg, whom I had known for years, was a clever man who in Himmler's name had established contact with certain neutrals and the Allies. But he had little experience of foreigners.

"Of course only very few people knew of these matters of high politics. Junior dyed-in-the-wool S.S. Groups just did what they were told. So I was not at all surprised when I had confirmation jrom another quarter that we were to be arrested at the very time when Hitler was giving Speer his last photograph—with a barely legible autograph.

"Governor-General Frank, coming from Prague and landing on our airfield at Kalten-Kirchen to join Dönitz, innocently gave the news to my Captain Eicholz during the drive to Eutin. He said he thought we had long been 'dealt with', to use his own tactful expression. An hour later he found himself in the presence of Speer and myself and I told him that he had just been flying with one of my crews, who could be relied on to return him safe and sound.

"At that time I had almost decided to make my escape. The aircraft stood ready to take off. We were supplied with everything we needed for six months. And then I found I could not do it. Could I bolt at the last moment, desert Germany and leave in the lurch men who had always stood by me? I must stay by my men.

"And now Himmler wanted to see me.

"As I am returning from a short visit to some Lübeck friends I drive past hordes of refugees on the roads and see wounded soldiers lying untended in the woods, mothers with newly-born babies cowering by the roadside, prisoners from the concentration camps being driven like cattle in the darkness. Roads of death.

"And God is silent?

"My mind is made up, all doubts and fears are banished. I will go to Himmler and tell him frankly what I think, as I always have. Until day dawns I sit up writing a letter to my wife and son. An officer is sent specially to Bavaria to find Galland.

"Then I drive off. It is the 28th April, 1945. I have been told that at the moment Himmler is at Güstrow in Mecklenburg. Here again the roads are blocked with refugees. There is no traffic going east. In the faces of the thousands of poor wretches whom I pass I can see that unspeakable misery which Dante tried to put into words in *La Grande Tristezza*. Here it has become flesh and blood.

"After five hours' driving we reach Güstrow. S.S. patrols show us the way. They had been told of our coming. My friend Knemeyer sits silent next to me. He has not deserted me. We drive to a country house. An S.S. sentry takes us into the house and tells us to make ourselves comfortable in the living room. The Reichsführer will send for us at once.

"I throw off my heavy cloak and look round. Everything here is simple and dignified and one feels at once that the furniture, pictures and lamps have been lovingly created by real artists. I cannot escape the contrast with overloaded Karinhall, but why it comes to mind I do not know. I mention it to my companion. Knemeyer smiles and points to two photographs in front of a mirror. I have a closer look and no one could have been more surprised. The two portraits are in silver frames. One is autographed as follows:

" 'Sir Henry Deterding—in the name of the German people, for your noble donation of a million reichsmarks.

Adolf Hitler.'

"Under the second photograph, which shows the Commander-in-Chief of the Luftwaffe in a mediaeval German stage costume holding a large hunting-knife, the astonished visitor read:

" 'To my dear Deterding, in gratitude for your noble gift of Rominten Reichs Hunting Lodge.

Your Hermann Göring.'

"Now we know where we are—at Krakow, in Sir Henry Deterding's Mecklenburg country place.

"The life of this great man of business passed before my eyes. A few days ago I happened to be talking to an old friend of Sir

Henry. And now I have come across these photographs with their inscriptions. I can hardly refrain from stuffing these two historical documents into my brief case to add to my collection to witness the fact that in this world only money really counts and Might is Right.

"Henry Deterding was knighted by King George V on New Year's Day, 1920, and a book about him, *The Victory of Shell over the German Empire*, appeared the same day. It was the same Deterding who built up his oil trust with the old Paris banking house of Rothschild and changed the name of his company from Royal Dutch to Shell. It was under the same Deterding, who was nicknamed 'Napoleon', that Shell became the main pillar of the British world empire. The *Entente* powers had swum to victory on the great streams of oil he poured out for them.

"Then he became the sworn foe of the U.S.S.R. and the 'red oil', which he had lost at Maikop, Grosny, Baku and in the Urals and Turkestan. Had he not once written that a permanent solution of the oil crisis would never be possible before there was a satisfactory solution of the Russian problem? Was that the reason why he had backed National Socialism, or was there some other explanation of these donations and charities? He alone could clear up all the political background. But Sir Henry Deterding is dead.

"And now the Deterding dream itself is about to fade away for ever. In a few days the Russians will be here to occupy the German country place of a Knight of the British Empire. Are the Russians stronger than National Socialism and Deterding put together?

"The voice of the A.D.C. breaks in on my musings: 'The Reichs-führer will receive you.'

"I am taken along a narrow corridor and up a winding staircase guarded by S.S. sentries to Himmler's study. The room seems to have been left as it was, but in one corner there is a machine-pistol within range. Its safety-catch is off. Himmler, in high black boots, field grey uniform of some coarse material and with the skull and S.S. insignia on the collar, is sitting at a desk alone. His sleeves are much too long and half conceal the hands, which have something uncanny about them. There is a forlorn, cheap ring on the little finger of his left hand. It must be a memory of the past. Himmler's face, which I have often studied, has nothing special about it. It looks unhealthy, somewhat puffy, wax coloured, and the chin is too small. Two cold, indifferent eyes size me up from behind pince-nez. He seems very overtired, but his greeting is emphatically friendly. His handshake is not that of a man of energy. Everything about him is curiously soft and he seems almost harmless. Is he not known as 'soft Heinrich'? And yet his impenetrable personality fills the room.

"He motions me to sit down. 'I've sent for you to clear up some Luftwaffe problems. In the very near future I must expect to be negotiating with our enemies, probably through some neutral country. The war has entered the final stage and there are some very important decisions I shall have to take. The Führer is isolated in Berlin. I shall be the only man to prevent chaos in Germany and I think that foreigners will not negotiate with anyone but myself.'

"Himmler speaks in his strong Bavarian accent and his voice is as level as if he were discussing something of no importance whatever. He enlarges on the possibility of continuing the war in Mecklenburg and Schleswig-Holstein. He will form Freikorps, as in the old days of the War of Liberation. The position is far from hopeless. Himmler talks as if he were Hitler's successor already—not in so many words, but with broad hints. Then he continues less enigmatically and I am at last told why I have been sent for: 'I've already told you that in the very near future I shall probably have to negotiate through a neutral country. I've heard that all aircraft suitable for the purpose are under your command. What possibilities are there?'

"I gazed out at the well-kept park through the broad windows. I knew about Himmler's vain peace-feelers to the West back in 1940 and others in the late autumn of 1944 through Switzerland; now there appeared to be an approach through Sweden. What was the 'Truest of the True' up to now? 'Reichsführer,' I answered after a pause, 'I was examining the map of the world yesterday to see where we could fly to. I have planes and flying-boats ready to fly to any point of the globe. The aircraft are manned by trustworthy crews. I have given instructions that nothing is to take off without a verbal order from myself.'

"Himmler's voice is a tone lower as he replies: 'I think you have misunderstood me. What I mean is that if I have to start negotiations I shall probably need some aeroplanes. Have you got some?'

" 'I have enough aircraft ready to start at any time for negotiations to be proceeded with.'

"Reassuming a more friendly tone, Himmler asks me where he can get hold of me in the next few days. He will probably need my help on the lines mentioned. I tell him he can find me at Travemünde airfield where part of my command is stationed. The A.D.C. comes in to announce the arrival of Field-Marshal Keitel. Himmler rises as a signal to leave. Then I suddenly go mad and, standing in the doorway, cannot resist a parting shot:

" 'Surely you're not going to ask military advice from Field-Marshal Keital?'

" 'What do you mean?' says Himmler, looking at me sharply.

"I give him a cool smile. 'I was only thinking that you cannot win wars with a Field-Marshal Keitel. But of course I'm only an airman and know nothing about it.'

"It was not the first time that I had spoken to Himmler in that strain.

" 'I'm very glad to have a military adviser like the Field-Marshal.'

"He is about to add something when I realize that Keitel is already coming in. An orderly rushes out after me to say that the Reichsführer has arranged for us to have something to eat before we return. Sandwiches and coffee are served. The S.S. guards give us a stiff salute as we pass through the barriers in the streets. Our return journey is harassed as usual by low-flying enemy aircraft. We pick up a badly wounded soldier and deposit him at Schwerin hospital. After driving for twelve hours I am back with my men. We are almost overdue.

"Shortly afterwards I am summoned to Eutin by Grand Admiral Dönitz. When I enter the naval barracks just outside the town late in the afternoon of the 30th April, that part of the Government which fled northwards, is gathered there. Himmler and Rosenberg are present.

" 'The Führer is dead. Long live the Führer!' "

Karl Dönitz has been appointed his successor. The old submarine commodore can never have dreamed that instead of directing his U-boat flotillas he would one day be steering the battered ship of state through the stormy waves of time. But why him? The main thing was that he was under orders. At the last hour Hitler had broadcast that his successor was to be Karl Dönitz. He had previously damned Göring, his "loyal paladin", to eternity. A soldier carried out orders. That was his training. Why should he not do the same as politician and statesman—to the best of his ability? And Karl Dönitz took his thankless task very seriously.

Far more important was it to the late ministers and men who were only too anxious to replace them. No one would admit that the Allies were simply looking for someone to sign an unconditional surrender. Speer called what came next an "operatic government", and to all appearances he was right. Yet legally Dönitz's government was the unquestioned successor to the Hitler régime. It believed it was recognized by the Americans and at first that was true. But if one surrenders unconditionally one must expect to be treated unconditionally afterwards. And that is what happened, to a degree defying imagination. How could we have seriously thought that the Western Allies would suddenly make common cause with the "Nazi generals and Nazi ministers", considering that only a short time previously Roosevelt had handed over Berlin and East Germany to his "great comrade" Stalin?

The following sketch of events stems from those mad days of universal confusion and dissolution:

"A hutted camp at the gates of the little town of Eutin in Holstein. Heavily armed U-boat men mount guard at the new 'Wolf's Bunker' (Führer Headquarters). Everything looks mighty warlike, and it is just as well as the enemy is at the gates of Hamburg— barely fifty miles away as the crow flies. All the heroic defenders of their Fatherland are here! Generals and Admirals at battalion strength and wandering about looking for suitable employment. A Luftwaffe general goes mad and thinks it would be a good idea to embody aircraftless pilots in a division with which 'Fortress Schleswig-Holstein' could be held for at least a year. Incidentally, we flyers have barely a revolver left and understand as much about land fighting as General Staff officers do about air fighting. The recognized grandees of the Third Reich have already exchanged their brilliant uniforms for quieter civilian clothes.

"Heinrich Himmler, wearing a heavy leather overcoat with the collar turned up, appears in a light-armoured car, its anti-tank rifles sticking out like asparagus shoots. He had counted on in-heriting the coveted post of 'Ersatzführer'. But he was loyal. He even said he did not want to remain Minister for the Interior in the new government. But he was never far away. Loyal—to whom? Be that as it may. Dönitz strengthened his U-boat bodyguard the same evening.

"After a few minor crises in the government's quarters we have newly-baked or warmed-up ministers, commanders-in-chief and secretaries of state again. Among them were Count Schwerin-Krosigk, Albert Speer, Keitel, Jodl, Stuckart and Ohlendorf. From time to time the Ribbentrops, Himmlers and Rosenbergs, all without portfolios, bobbed up.

"What was to be done? That was the vital question. The first thing was to 'get another move on', as the British were in Lübeck. No reinforced battalion of generals and admirals made a move against them. This last period was conspicuous for the absence of the heroic, of which there was so much talk and for which vast numbers of men on all fronts were dying. And yet it would have been so easy: *dulce et decorum est pro patria mori.*

"The other alternative was chosen and under the umbrella of a rainy, foggy evening, a procession of cars, arranged according to seniority, crossed the Kaiser Wilhelm Canal and made for Flens-burg. There a heartening spring sun greeted us and the fjord was already beginning to bloom. Obviously nature was not worrying about the troubles of the newly-arrived guests who made their quarters in the Naval Training School at Mürwik. When the British came on the scene—which was very soon—they too paid no particular attention to the 'Reichs Government' which was losing touch with reality in hour-long cabinet meetings and other official business. All that the British were waiting for was orders.

"Speer was still the outsider. He took up residence in the romantic Glücksburg and only went occasionally to the 'liquidation head-quarters' at Mürwik.

"Meanwhile unconditional surrender had been duly signed at Rheims and proclaimed in Berlin without making any change in the ordering of affairs at Flensburg. We were beginning to raise our eyebrows over this idyll, which was no doubt due to some failure of stage-management on the part of the Allies. But all our doubts were soon set at rest.

"The 'Desert Rats', a famous British division, arrived one night and next day the Dönitz government had ceased to exist. Only one shot was fired—by an old sailor, Admiral von Friedeburg, who committed suicide. On the same day the men at the top were flown to Mondorf in Luxemburg. We stood at the edge of the tarmac and watched the foreign aircraft lose themselves in the haze on the horizon and disappear from sight. They left a certain amount of dust over the runway which took some time in clearing. It was the last flicker of the Third Reich. The curtain had fallen.

"We stood in silence, facing an uncertain fate, and quietly dispersed, no longer hopeful youth or cheering mob or just blind followers. Yet in spite of everything we dare not despair, however easy to do so. 'The virtue of a conquered nation is patience, not resignation', writes Oswald Spengler. And we are all such sticklers for virtue."

Then we were all herded behind barbed wire, and with us millions of German soldiers. Before long the whole of Germany was a giant prisoner-of-war camp. The victors had the power and they used it to the limit. For months hundreds of thousands of men slept on the bare ground crowded together like cattle with not very much to eat. Was that the democracy we had been promised?

German rights, international law—what did they count for now? In the West as in the East German property was war booty. The Germans saw themselves stamped as an inferior sort of human being. Montgomery forbade his men to play with German children. *No fraternization!* Ideals of humanity cherished by many Germans and preserved from times long past were faced with another and worse crisis.

In Glücksburg I was openly robbed of everything by men of the Cheshire Regiment, in the 15th Infantry Brigade.[1] I also had the opportunity to observe similar shameless behaviour, such as the theft of jewellery from the family of the Duke of Schleswig-Holstein and the last Grand Duke of Mecklenburg. They were both closely related to the British Royal Family.

In conversations with senior Allied officers I missed no opportunity of drawing their attention to the incalculable consequences of such misbehaviour. I met men, usually fighting men or airmen (the British Group Captain Felkin and the American General Banfill for instance) who entirely agreed with me. They had taken appropriate steps in their own commands but confessed that there was not much they could do to change the general situation.

In many respects I was glad when I was flown to England with a group of German engineers and scientists. There I met many people I knew. In Wimbledon were many scientists who were being "pumped". Latimer camp was full of Luftwaffe generals and general staff officers undergoing interrogation. Many hopes were buried there and elsewhere. "Unconditional surrender" was the motto. Only those whose intellects, inventiveness or technical knowledge were of really high interest were given an "agreement". In those cases it was immaterial whether the expert concerned was classified as a "war criminal", economic slave-driver, militarist or Nazi. The man himself was quite unimportant; it was his inventions and his brains which counted. If he was wanted he could go to America or stay in England. There was a strong demand for the inventors of the V-weapons, guided bombs, the new U-boat power units and jet aircraft.

On the English testing aerodrome at Farnborough could be seen captured German aircraft, including the Me 262, Me 163 and other types. The British engineers candidly admitted that most similar English aircraft were inferior.

In London considerable areas, particularly in the vicinity of the docks, were no more than waste spaces as the result of German bombing in 1940–41. The British Museum was still without windows. The damage done by V-bombs was less obvious, as the blast area was less.

In the courtyard near Westminster Abbey we saw British corporals training schoolboys. Only a short time before a British officer had been holding forth about ruinous German militarism. In London food was short and rationed. To the outward eye Britain did not

[1] In 1949 Scotland Yard recovered from an English soldier the 'Air Commander Führer decoration with diamonds' which had been stolen from me. The British Government returned it to me with an expression of its extreme regret at what had happened and the assurance that the culprit would be severely punished.

look like a victor. South of London I saw the remains of the primitive defences and trenches which had been produced in anticipation of a German invasion in 1940. Conversation with the English confirmed one's suspicions that the country was in no way armed or equipped for the extension of the war to the British Isles. There was even a shortage of rifles at that time.

Six months later I returned to Germany and was released from captivity. It was the year of the war criminal trials. At the Tegernsee country house of my dead Luftwaffe friend Harry Rother I met the Reuter correspondent, Mr. Papel Hamsher, who had been assigned to the international military tribunal at Nüremberg. What he told me compelled me to leave for that city at once in order to offer myself as a witness on behalf of my friend Albert Speer.

Nüremberg, the old imperial city, was no better than a heap of ruins. Its mediaeval atmosphere had departed under a hail of Allied bombs. But the Law Courts had survived. The guard was supplied by American military police. An enormous staff of Allied legal personnel and journalists, German defence lawyers and super-numeraries, gave the immediate vicinity of the Law Courts a special character. The inhabitants of Nüremberg paid little attention to the trials. They had too many worries of their own.

The arrangements for custody and supervision were extra-ordinarily strict. Fear of possible incidents seems to have been the reason. But the guardians slipped up from time to time.

Göring unquestionably returned to his old form during the trial. Once more he was the "Man of Iron". For months he and the other accused stood up to the daily tribulations of such a trial with dignity.

It seems to me premature to say more about this trial here or to describe events which would carry me beyond the limits of this book.

One or two letters from my personal correspondence with Albert Speer may reveal something of the human side. There is no point in wasting words on the legal aspect. I could not be more critical than General Fuller.

Nüremberg.
30th July, 1946.

My Dear Win,

In my cell here I have been thinking so much about all we have been through together and it has given me so much pleasure that it goes against the grain to address you as anything but "Du".[1] I know you will feel the same.

Here is something to cheer you up; I'm still just what I was in the

---

[1] 'Thou' – only used between relatives and very great friends. Of course I have to substitute 'you' here. Baumbach and Speer are not Quakers! (Tr.)

spring of '45. I hope to come through worthily. It helps me to know that part of the nation will always hold my name in honour. Nothing else matters. In those days I did what I did without regard to what might happen to me. It must not be different now. I am sure that I will get through this last stage with decency and dignity.

The task you have set yourself is splendid—to tell the truth about the consequences of the régime. The final phase will itself show the nation and the world what *fundamental* dangers to the future of society such régimes represent. It must be made easier for the German nation, which clings so tenaciously to the past, to free itself in heart and mind from the old ways.

I am sure that in this very moment of national disaster the German nation will develop life-giving forces. It may be that they will produce intellectual and artistic achievements which will in some measure compensate for our evil reputation in the world in the last few years.

Where is Napoleon now—and how much more does Beethoven mean to the world today!

I have found an eloquent passage in Hölderlin here: "The heart's breakers would never surge so splendidly if they had not to dash themselves against that silent old rock—fate." (Hyperion 1, 1) Perhaps the German nation will find itself purified while the world outside rushes into technology and the pursuit of pleasure.

The main thing is that we should all be true to ourselves and hold up our heads. I am finding that the last months of the war were harder to bear than anything that can come now. . . .

A.S.

Virchowstrasse 19,
Midnight, 30-31 July, 1946.

Dear Albert,

It is midnight and I am sitting at an open window listening to the crickets in the high grass outside. In this quiet hour my thoughts are yours, and I find it hard to put them into such arid things as words.

This year has enabled me to see my way at last. With our eyes on the future of those who come after us, it must be our endeavour to tell them the truth about an era and its actors which we personally have survived. Have I then failed to keep faith? Am I a deserter?

I have had my eyes open and can recall many experiences and impressions of the recent past and weigh them up. I can link them up and say—here is the answer!

When I do so they show that I am a changed man. I cannot remain for ever bound by any decision I made as little more than a child under the influence of schoolmasters and service superiors. It is a man's right to live, grow and mature.

But if the ideals of one's youth dissolve into thin air, is one breaking faith? To what is the loyalty owing—to those ideals I cherished as a boy in a boy's dreams? To prescribed ideas and programmes I adopted before I even knew what they meant? Or to the men with whom I threw in my lot who had not got it in them to live and grow but could only stay at the point where they stopped? No reasonable man can reproach me for disowning such loyalty. I must do what I have to do, even if it means being disloyal. . . .

He who steadily seeks truth and nothing but the truth will live by the truth, and the whole truth means the power to do apparent impossibilities. We want to be purified from all doubt. Our plea and prayer is: may our faith never cease to grow.

Columbus had that faith when he insisted on the existence of the New World. It exists in all who believe in their star and represents a true living force in man which in some way unknown to us bears him to his goal.

As we came from Nothing, so shall we return to Nothing—we who thought we were the centre of the world and that our end would mean its end also. The veil of death hides God, in Whose image we are created, and Who alone gives us hope that we may share the fullness of His Being. God and man can come together at any time and in any place. There is only one foundation for Germany's hope—that God will help us

Yours,

Werner.

On the evening, a few days before the verdicts, on which I was driving out of the city, with the well-known defence counsel, Dr. Dix, and Göring's counsel, the Hamburg lawyer Siemers, I asked Herr Dix what the outcome was likely to be. "Herr Baumbach," he replied, "you know the answer as well as I do. It's a pure gamble. Most of the accused will be hanged, a few will get long sentences and one or two may possibly be acquitted. I can't say which will be which." I had to rest content with this significant answer from a leading lawyer.

Speer, who preserved his honesty and maintained his human dignity even in this predicament, had this to say of those representatives of the victorious powers who sat in judgement on the vanquished:

". . . May this trial contribute to the prevention of unnatural wars in the future and lay the foundations of human co-existence. After all that has happened, what does my own fate matter in comparison with that high endeavour?"

Did Nüremberg and the many other "War Criminal Trials" have that effect? Having regard to the developments in recent years, that question can confidently be answered in the negative. The victors themselves have not adhered to the *ex post facto* rules they laid down at Nüremberg. It is already clear that Nüremberg was no beginning of an era of peace and justice, but the seedbed of much greater evils—to make war a crime for *everyone*. The noose is waiting for the loser. So stop at nothing in war lest you become the loser!

CHAPTER XX

TOO LATE?

Too late? After the bitter experiences of the Second World War this question presented itself spontaneously to many thinking men, among the victors as well as the vanquished. Deep within the hearts of the nations ravaged by the war slumber a sense of horror and a secret feeling that a new catastrophe would be on an incalculable scale and have unprecedented consequences.

Even the Second World War has not been able to solve the burning problems. It has led not to a decision but a general indecision. It brought no end to crises—not even in Germany, whose defeat has culminated in an outbreak of international political disease affecting the world's most sensitive spot. Only its treatment and cure can save us from ending up in the mass graves of the nihilist inferno.

The last war can be considered only as a transition to a new era of humanity with completely new conditions of life. It was not the crisis itself but only one, and perhaps not the decisive, phase. In my view we have not yet reached the peak of this turning-point. At this stage in the evolution of history it is already difficult to judge of recent events *sine ira et studio*, and I think it is all the more difficult to arrive at a comforting prognosis from the present fluid conditions.

So far little has happened to save us from a third world war. More than fifty so-called peace conferences have not been able to give optimistic humanity its much desired peace but simply created a general feeling of insecurity which is paralysing reconstruction. Even today mistrust, if not downright hatred, prevails between the nations instead of the mutual confidence which is requisite for peaceful co-operation. Freedom, for which this second world war was also fought, has been like melting snow in the hot sunshine to many nations, and others must be wondering how long they can preserve their own.

To add to all this, we are progressively becoming slaves to science, science regarded as an end in itself. Looking at the failures due to our own shortcomings we help ourselves out with mindless masses of steel and concrete. As never before Man, the helpless weakling, worships the power of matter. Even a statesman like Churchill sees power, in the shape of the atomic bomb, as the only sure basis of freedom. But this potential power has not prevented offensive war in China and elsewhere, and it seems to me that even in Europe men are beginning to wonder whether naked power is a sound foundation for peace.

After the 1914-18 war the first great onslaught of Bolshevism failed. European civilization still possessed enough confidence and vigour to resist it. But will this always be the same after the Second World War? The amalgamation of German and French industry, an economic alliance of the Western European states, including Germany, are a desirable but modest beginning to the unity of the culture nations of Europe; but they have not yet established a really strong defensive barrier against the Soviets. In this connection it should be remembered that West Germany, artificially over-populated as she is, can only exist economically and avoid another financial catastrophe if allowed a greatly increased export trade and her exports are admitted to other countries. And the men at the top in high politics must always bear in mind that it depends on that decision whether democracy is finally divorced from dictatorship or not.

As long as war against Russia is possible, the leading Western Powers concerned with the military problem must not overlook two factors of great importance—the psychological and political wars. In those spheres the Soviets have for long left no stone unturned. Most vital of all is a peace programme. It must have long term aims. No Atlantic Charter, but a *pax americana* with a *pax christiana* as its social content!

It may be that only the Russians themselves can put an end to Bolshevism, as it is a Russian disease. But as a danger to humanity it can only be rooted out by a spiritual conquest of atheism and nihilism. These two are the pillars of aggressive power politics. Generally speaking, the recognized Christian religions have put up only a divided and unsystematic resistance to them. Judging by past experience, nihilism cannot be conquered solely by an economic system (socialism) or by a spiritual force (religion), but only by a complete penetration of our whole moral, intellectual and economic life by the spirit of Christ.

Will the West be capable of assimilating such a novel, awe-inspiring idea? It is certain that the next ten years will establish the mental and moral complexion of the world, for a long time to come.

I must refer briefly to the mighty changes with which our world is faced.

The air above the continents and oceans is wide open. It was once freer than the sea. Then man's first attacks upon it spurred on science and technology to devise rules for its control. Politics and economics were not slow to seize on the infant aviation and impose their requirements on it. Man's insatiable curiosity and the spur of sporting rivalry led to a hard and largely state-aided battle for the air routes of the world. Then the international struggle for the command of the air, bitter enough in peace, became a matter of life and death in war.

Thereupon, previous notions of time and space called for rectification. The traditional laws and ideas of war, the so-called classical rules of strategy, ceased to apply. The world has become smaller. The aeroplane knows nothing of frontiers and natural barriers. It can seek and find its way over sea and land to the most distant target in any weather. This revolutionary means of transport opens up enormous possibilities both for commerce in peace and strategy in war.

The power of modern states is now based mainly on their air potential. Just as history has shown that in the past a great power must be either a sea power or a land power, in future the nations who wish to keep their place in the sun must arm themselves with air power.

Of course the land will still continue to be man's home and the sea will continue to be a traffic highway, showering us with food-stuffs like dew. But the land will no longer merely provide the meeting-points of ocean and land traffic routes, which will be supplemented, intersected and even replaced by air routes. There will be new air traffic centres quite independent of the world's traditional trade routes.

Flying over the Poles will soon be commonplace. The continents are getting closer to each other. Land bridges and sea routes once important are being outflanked. The Arctic regions seem destined to be the world's traffic centres of the future.

No known period of history has been so essentially dynamic. Science and technology are rivals in a mighty competition and man seeks in vain to preserve the balance with their most recent creations, only to find himself overtaken again and again. The struggle for space beyond the reach of the earth's gravity has already begun. Aircraft flying faster than sound have penetrated to the stratosphere. Flying robots, guided from a point on land, will soon leave the gravitational area. Research workers are talking of the fourth dimension and final infinity. Air travel will be replaced by space travel.

The developments both from the angle of geopolitics and air warfare in the second half of a century strewn with crises may well take the following course:

The continents and oceans will be linked up by the shortest routes via the "Arctic" central air ports. Alaska, Greenland, Iceland, Scandinavia and northern Siberia will be the immediate neighbours of this giant base. The Arctic air front, with its barrier zones and airfields, will develop both sides of the 20th and 160th meridians respectively. America on one side; Northern Europe forms the critical flank. He who holds it protects its strategic moves, whether to or from the East. The Atlantic becomes a sort of Mediterranean, an inland sea between its eastern and western neighbours. Europe is the half-way house on the route between east and west. It is both a protection and a periphery. But, like Africa, in itself it is a torso.

In the coming age of space travel and intercontinental air strategy Russia, with her Janus head turned both in the Atlantic and the Indo-Pacific direction, can no longer maintain her isolation or use her space as a weapon as heretofore. If she continues to remain outside the natural community of nations, with the selfish object of inciting their proletariat to look to Moscow and stage a world revolution, she will necessarily compel the rest of the world to defend itself. From the Arctic base its new Siberian armament can be reached directly by air. From the Anglo-American air bases in the Far East, the Indo-Pacific area, the Middle East, the Mediterranean, Alaska, Greenland and Iceland all the known and prospective Russian military and supply centres are exposed to attack by superior air force, even in the present stage of aviation technique. A strategic air blockade of the Eurasian land mass can be established at any moment.

The potentates in the Kremlin know the possibilities of the policy of encirclement and are trying to gain time for two reasons: to catch up with, and if possible overtake, the Anglo-American war potential, both in quality and quantity, and to extend the area under Russian control as much as possible in order to reduce vulnerability to air attack to a minimum. The systematic decentralization of their industries ￼and the creation of regional defence zones in the areas outside Russia proper all point in that direction.

Man stands at the crossways. It lies in our power to guide and control the material forces which the human mind has released or to let them get out of hand.

For years I have been witnessing the efforts of the victorious powers to re-educate Germany. They were not calculated to realize their purpose. Sins of commission and omission made them fail to convince. I cannot help thinking that they would have been better

14

advised to learn from the Germans the sad end of such dangerous crusades.

The Europeans, the Western nations, are now cooped up together in the white man's Noah's Ark—and all hands are needed in a storm. But the days of galley-slaves are past, and the Allies must realize it if they appeal—somewhat late—to the younger generation in Germany. Re-militarizing those who have been demilitarized is not quite such a simple matter. Their own front-line fighters could be the best bridge to reconciliation and rehabilitation.

In the fiery trial of this turning-point in history, Germany has one great advantage, a moral one, over the rest of the world. In his hour of greatest need, the German can get up on his own feet again. The great catastrophe has taught our people how to distinguish between the material and the immaterial, the true and the false, the temporary and the eternal.

We too are at the parting of the ways. Once more fate has snatched us back on the edge of the abyss. If we learn our lesson we have lost much but we shall be back on our own feet. If the rest of the world does not heed the warning given by the disaster to Germany it will have to bear the inevitable consequence. It will then have to be said that the moral and intellectual atomization of man preceded the physical atomization of the globe. In that process air power is but an auxiliary. It is a part of human activities and it depends on us to guide it into constructive, not destructive paths.

If we now fail to control not only man and his work but the unleashed forces of nature, a generation born too late whose feeble hands cannot master its own fate will see the dreadful *mene tekel* of all vanished nations and lost glories—

TOO LATE!

APPENDIX

	at 1st Nov. 1943	1st Jan. 1944	1st June 1944	15th Feb. 1945
N.C.O.s and men ..	1,970,000	1,732,000	1,809,000	1,331,700
Officers, officials, engineers	119,000	120,000	120,000	120,000
Auxiliaries, including foreigners	430,000	450,000	510,000	546,000
Civilians	475,000	480,000	450,000	201,000
Totals ..	2,994,000	2,782,000	2,889,600	2,198,700

of which at 14th Jan., 1945, 26,411 officers and 632,486 N.C.O.s and other personnel were trained airmen.

CASUALTIES AMONG GERMAN TRAINED AIRMEN BETWEEN 1ST SEPT., 1939 AND 28TH FEB., 1945:

Killed	44,065	(6,631 officers)
Wounded and injured	28,200	(4,245 officers)
Prisoners and missing	27,610	(4,408 officers)

GERMAN AIRCRAFT PRODUCTION 1914–18

1914	.. ..	1,348 aircraft of all types
1915	.. ..	4,532 ,, ,, ,, ,,
1916	.. ..	8,182 ,, ,, ,, ,,
1917	.. ..	19,716 ,, ,, ,, ,,
1918	.. ..	14,310 ,, ,, ,, ,,

TOTAL OUTPUT OF GERMAN AIRCRAFT 1939–45

	1939	1940	1941	1942	1943	1944	1945	Total
Bombers ..	737	2,852	3,373	4,337	4,649	2,287	—	18,235
Fighters ..	605	2,746	3,744	5,515	10,898	25,285	4,936	53,729
Battle aircraft	134	603	507	1,249	3,266	5,496	1,104	12,539
Reconnaissance aircraft ..	163	971	1,079	1,067	1,117	1,686	216	6,299
Transport aircraft	145	388	502	573	1,028	443	—	3,079
Naval aircraft	100	269	183	238	259	141	—	1,190
Gliders ..	—	378	1,461	745	442	111	8	3,145
L. of C. aircraft	46	170	431	607	874	410	11	2,549
Training aircraft	588	1,870	1,121	1,078	2,274	3,693	318	10,942
Jet aircraft ..	—	—	—	—	—	1,041	947	1,988
Totals	2,518	10,247	12,401	15,409	24,807	40,593	7,540	113,515

Further Statistics:

BOMBERS

1. New types from 1939 to 1945 which went into front-line service, 2,637 aircraft.
2. Modifications and production of types already in service in 1939–40, 27,773 aircraft.

FIGHTERS

1. New types from 1939 to 1945 which went into front-line service, 21,025 aircraft.
2. Modifications and production of types already in service in 1939–40, 36,335 aircraft.

EXPENDITURE ON RESEARCH, DEVELOPMENT AND PRODUCTION OF AIRCRAFT

	approx. million reichsmarks		approx. million reichsmarks
1935	400	1940	5,000
1936	900	1941	6,500
1937	1,000	1942	9,000
1938	1,000	1943	12,000
1939	3,000	1944	12,000

German aircraft production by type and purpose (not including prototype aircraft):

1. BOMBERS

Type	1939	1940	1941	1942	1943	1944	1945	Total
Ju 88 ..	69	1,816	2,146	2,270	2,160	661	—	9,122
He 111 ..	452	756	950	1,337	1,405	756	—	5,656
Do 17 ..	215	260	—	—	—	—	—	475
Do 217 ..	1	20	277	564	504	—	—	1,366
He 177 ..	—	—	—	166	415	565	—	1,146
Ju 188 ..	—	—	—	165	301	—	—	466
Ju 388 ..	—	—	—	—	4	—	—	4
Totals	737	2,852	3,373	4,502	4,789	1,982	—	18,235

2. FIGHTERS (MAIN TYPES INCLUDING NIGHT FIGHTERS)

Type	1939	1940	1941	1942	1943	1944	1945	Total
Ju 88 ..	—	62	66	207	406	2,518	355	3,964
Do 17 ..	—	9	—	—	—	—	—	9
Me 110 ..	156	1,008	594	501	641	128	—	3,028
Me 109 ..	449	1,667	2,764	2,657	6,013	12,807	2,798	29,155
Me 210 ..	—	—	92	93	89	74	—	348
Fw 190 ..	—	—	228	1,850	2,171	7,488	1,630	13,367
Me 110 ..	—	—	—	—	789	1,397	54	2,240
					Total of Fighters			53,729

3. JET AIRCRAFT

Type		1939	1940	1941	1942	1943	1944	1945	Total
Ar 234	..	—	—	—	—	—	150	64	214
Me 262	..	—	—	—	—	—	564	730	1,294
Me 163	..	—	—	—	—	—	327	37	364
He 162	..	—	—	—	—	—	—	116	116
						Totals	1,041	947	1,988

4. GROUND ATTACK AIRCRAFT

Type		1939	1940	1941	1942	1943	1944	1945	Total
Ju 87	..	134	603	500	960	1,672	1,012	—	4,881
Hs 129	..	—	—	7	221	411	202	—	841
Fw 190	..	—	—	—	68	1,183	4,279	1,104	6,634
Ju 88	..	—	—	—	—	—	3	—	3
Totals		134	603	507	1,249	3,266	5,496	1,104	12,359

5. RECONNAISSANCE AIRCRAFT

Type		1939	1940	1941	1942	1943	1944	1945	Total
Ju 88	..	—	330	568	567	394	52	—	1,911
Do 17	..	16	—	—	—	—	—	—	22
Do 215	..	3	92	6	—	—	—	—	101
Fw 200	..	1	36	58	84	76	8	—	263
Hs 126	..	137	368	5	—	—	—	—	510
Fw 189	..	6	38	250	327	208	17	—	846
Me 110	..	—	75	190	79	150	—	—	494
Me 109	..	—	26	—	8	141	979	171	1,325
Me 210	..	—	—	2	2	—	—	—	4
Ju 188	..	—	—	—	—	105	432	33	570
Ju 290	..	—	—	—	—	23	18	—	41
Me 410	..	—	—	—	—	20	93	—	113
Ju 388	..	—	—	—	—	—	87	12	99
Totals		163	971	1,079	1,067	1,117	1,686	216	6,299

6. SEAPLANES AND FLYING-BOATS

Type		1939	1940	1941	1942	1943	1944	1945	Total
Ar 196	..	22	104	94	107	104	—	—	435
Do 18	..	22	49	—	—	—	—	—	71
Bv 138	..	—	39	82	85	70	—	—	276
He 115	..	52	76	—	—	—	141	—	269
Do 24	..	—	1	7	46	81	—	—	135
Bv 222	..	—	—	—	—	4	—	—	4
Totals		100	269	183	238	259	141	—	1,190

7. TRANSPORT AIRCRAFT

Type		1939	1940	1941	1942	1943	1944	1945	Total
Ju 52	..	145	388	502	503	887	379	—	2,804
Me 323	..	—	—	—	27	140	34	—	201
Go 244	..	—	—	—	43	—	—	—	43
Ju 352	..	—	—	—	—	1	30	—	31
Totals		145	388	502	573	1,028	443	—	3,079

8. TOWED GLIDERS

Type	1939	1940	1941	1942	1943	1944	1945	Total
	—	378	1,461	745	442	111	8	3,145

9. LIAISON AIRCRAFT

Type		1939	1940	1941	1942	1943	1944	1945	Total
Fi 156	..	46	170	431	607	874	410	11	2,549

10. TRAINERS

Type	1939	1940	1941	1942	1943	1944	1945	Total
	588	1,870	1,725	1,078	2,274	3,693	318	11,546

BOMBERS—NEW TYPES

Type	1939	1940	1941	1942	1943	1944	1945
He 177	—	—	—	166	415	565	—
Ju 188	—	—	—	—	270	733	33
Me 210	—	—	94	95	89	74	—
Ju 388	—	—	—	—	—	91	12
Totals	—	—	94	261	774	1,463	45

BOMBERS ALREADY IN SERVICE IN AND BEFORE 1939–40

Type	1939	1940	1941	1942	1943	1944	1945
Ju 88	69	2,208	2,780	3,094	3,260	3,234	355
He 111	452	756	950	1,337	1,405	756	—
Do 17	231	275	—	—	—	—	—
Do 217	1	20	277	721	711	—	—
Ju 87	134	603	500	960	1,672	1,012	—
Totals	887	3,862	4,507	6,112	7,048	5,002	355

PISTON-ENGINED FIGHTERS (NEW TYPES DURING THE WAR)

Type	1939	1940	1941	1942	1943	1944	1945
Fw 190	—	—	226	1,918	3,354	11,767	2,734
Me 410	—	—	—	—	291	722	—
Do 335	—	—	—	—	—	7	4
Totals	—	—	226	1,918	3,645	12,496	2,738

FIGHTERS ALREADY IN SERVICE IN AND BEFORE 1939–40

Type	1939	1940	1941	1942	1943	1944	1945
Me 109	449	1,693	2,764	2,665	6,247	13,786	2,968
Me 110	156	1,083	784	580	1,580	1,525	45
Totals	605	2,776	3,548	3,245	7,827	15,311	3,014

INDEX

A.A. (flak), 25, 34, 44, 142, 148–53, 178
A.A. (British), 31, 81, 90, 178
Abdullah, Emir, 101
Aerial mining, 81, 85–7
Africa, North, operations in, 101, 102, 110, 132–41
Air Corps, II, VIII, X, XI, 99, 122, 135, 136, 137
Aircraft engines (German), 48, 62, 63, 83, 109, 161, 162
Aircraft production (German), 161, 162, 211–15
Aircraft types (American), 63, 162
Aircraft types (British), 50
Aircraft types (German)—
general, 54, 55, 211, 212
specific:
Arado, 33, 34, 47
Dornier, 19, 20, 33, 34, 47, 50, 59, 61, 63, 64, 93, 96, 162
Focke Wulf, 22, 47, 63, 93, 109, 160, 161, 212–15
Heinkel, 20, 33, 34, 45, 47, 53, 59, 61, 62, 69, 93–6, 105–9, 212–15
Henschel, 33, 94
Junkers, 19, 20, 21, 22, 33, 44, 47, 50, 53, 61–3, 69, 93, 94, 108, 109, 156, 162, 212–215
Messerschmitt, 28, 33, 45, 47, 63–7, 80, 109, 110, 156, 160, 165, 167–72, 212–15
Aircraft types (simplification of), 48, 49
Air Defence Zone West, 155
Air Fleet, Second, 135, 142

Air Forces:
Belgium, 31
France, 30, 31, 33
Great Britain, 30, 31, 33 (and see Royal Air Force)
Italy, 32, 33
Netherlands, 31
Poland, 32
Russia (Soviet Union), 32, 112, 113, 117, 120, 159
U.S.A., 30, 31, 40, 46, 157, 158
Air Ministry (German), 21, 22, 23, 26, 43, 47, 105, 165
Air-sea co-operation, 83–5, 107
Air Staff (see High Command, Luftwaffe)
Air Warfare School (Gatow), 57
Alexander, General, 141
Alexandria, 132
Algeria, 135, 136
Anderson, Major-General, 172
Anglo-American bombing offensive, 52, 53, 54, 65, 104, 111, 126, 153, 155–9
Anzio, 143
Antwerp, 179, 180
Appleton, 79
Archangel, 88, 89
Arctic Circle, 90
"Areopagus" Conference, 66
Armament Commissioners, 55
Armaments and Munitions, Ministry of, 44, 54, 148, 176
Arnold, General, 104
Aschenbrenner, Major-General, 112
Atlantic, Battle of the, 83–98
Atlantic Wall, 144, 145, 177
Augusta, 142